P9-AQH-716

# Augmented Learning

# Augmented Learning

Research and Design of Mobile Educational Games

Eric Klopfer

The MIT Press
Cambridge, Massachusetts
London, England

© 2008 Massachusetts Institute of Technology

All rights reserved. No part of this book may be reproduced in any form by any electronic or mechanical means (including photocopying, recording, or information storage and retrieval) without permission in writing from the publisher.

For information about special quantity discounts, please email ⟨special_sales@ mitpress.mit.edu⟩.

This book was set in Stone Serif and Stone Sans on 3B2 by Asco Typesetters, Hong Kong.
Printed and bound in the United States of America.

Library of Congress Cataloging-in-Publication Data

Klopfer, Eric.
Augmented learning : research and design of mobile educational games / Eric Klopfer.
    p.  cm.
Includes bibliographical references and index.
ISBN 978-0-262-11315-1 (hardcover : alk. paper)
1. Educational games—Data processing. 2. Educational games—Design and construction. 3. Simulation games in education—Design and construction.
4. Mobile computing. 5. Pocket computers—Programming. I. Title.
LB1029.G3K585   2008
371.33'7—dc22                                                    2007032260

10  9  8  7  6  5  4  3  2  1

# Contents

# Preface

For as long as there has been formal education, there have been calls for education reform. Sometimes those calls arise from emerging pedagogies. Sometimes they emerge from fundamental social and political change. And sometimes they come out of necessity. We are currently in a transition of necessity. Technological advances, globalization, and the shift from an industrial to an information economy are forcing our hand and demanding change. The rapid pace of change, on the order of years instead of decades, further demands that education and training cannot end after high school or college, but must become a lifelong pursuit.

This wave of change has brought with it a demand for new tools and approaches that can address the new learning demands of elementary, secondary, university, and lifelong education. E-learning has emerged as a serious contender to help support the learning needs of individuals across this spectrum of levels. E-learning itself can mean many things to many people and at its core simply means electronically supported learning, which can be online, on desktop PCs, or even on mobile devices (though the latter is sometimes referred to as m-learning). In practice e-learning often means delivery of information and content to learners through online hypertext, accompanied by images, audio, and video. But e-learning can mean much more, as evidenced by the recent surge of interest in using video games to teach everything from basic math skills for young learners to advanced communication skills for adults.

Researchers have shown that the potential of video games in education resonates with researchers, teachers, and students alike. There is growing interest in utilizing games as educational tools (Prensky 2001; Gee 2003). People are seeing the value of games as models of pedagogically rich, highly motivating learning environments. Commercial games possess many of the elements we look for in learning environments—collaboration, problem solving, higher-order thinking skills, and so on. While some "edutainment" games such as SimCity are being repurposed for the classroom, another set of "serious games" is being developed expressly for learning.

Yet the introduction of video games into the classroom, or even education more broadly, has yielded mixed results at best. Researchers worry that the skills learned in video games may be difficult to transfer to the real world. Developers of games lament the skyrocketing production costs and questionable market. Teachers fret about the practical difficulties of implementing games in the classroom. Students resent drill-and-practice activities disguised as games. The potential of video games for learning is real, but so are the difficulties in realizing this potential.

Enter mobile learning games. The mobile games market (encompassing cell phone games and handheld console games like the Sony PSP and Nintendo DS) is the fastest growing gaming sector in the world. Commercial industry trends, while well ahead of the educational sphere, demonstrate tremendous growth in the area of portable, handheld games. Nintendo Game Boy, Nintendo DS, Sony PSP, and many cell phone games are quickly becoming the fastest growing sector in games (Reuters July 28, 2005). While some of these games are merely translations of "big screen" games resized for smaller display screens, some of the most successful handheld games are written to take advantage of the unique aspects of these platforms. Mario Kart allows players to form ad hoc races with players standing nearby, while Nintendogs, a wildly popular game in Japan that is a more sophisticated version of virtual pets, requires players to "pet" their dogs using the Nintendo DS touch screen.

Handheld games are accessible to many more people than PC or console games, and can be developed without the astronomical production costs associated with blockbuster games. With more devices in the hands of potential players, a platform more amenable to casual play, and no need for expensive 3D high-definition graphics, this is a fast-moving sector. Yet the potential of handheld devices as a platform for learning games has gone essentially untapped. Handheld computers have incredible potential for aiding learning in a time when people must tackle complex problems and acquire information in "just-in-time" fashion. These portable connected computers can provide just the information necessary when it is needed and where it is needed. Their design, size, and mobility also make them an ideal platform for learning games. Rather than cramming desktop applications onto these small devices, such games play to the strengths of this platform—its portability, context sensitivity, connectivity, and ubiquity. Well-designed mobile games can use the physical and social context of the player as integral components to the game, creating a rich playing and learning environment.

This book describes the educational and gaming landscapes that place mobile learning games in the unique position to make a substantial impact on education. It starts by exploring the past and present of education, educational technology, edutainment, and mobile games, and then delves into case studies of a number of mobile educational games that have been designed, developed, and implemented in recent years. These case studies detail the history, uses, and design principles primarily of two categories of mobile educational games: participatory (requiring interactions with other players) and augmented reality (responding to player location) handheld simulation games. These games can place learners in real-world contexts that promote transfer of learning from one context to another. They can be produced at much lower costs, using social dynamics and real-world contexts to enhance game play. They can be integrated into the natural flow of instruction much more seamlessly than their big-screen counterparts. And they can create compelling and fun educational environments for learners. All of these factors combine to position mobile games in a unique and powerful position within the educational technology space.

## Why Mobility Matters: Theory and Practice

There is an adage in the field of learning sciences that we should define a problem first and then seek the most suitable technology to fix that problem, rather than seek a problem to fix with a particular technology. I'd be lying if I said that this was strictly how we conduct our work in the research and development of mobile learning games, here in the MIT Teacher Education Program or in the work to which I have contributed in the Education Arcade. To be honest, most research (including ours) is much more a hybrid of the two approaches. Surely one can find value in not strictly limiting the genesis of innovations to addressing an existing problem or challenge. Similarly, science is not conducted by the linear process of sitting down in a chair and coming up with a hypothesis to test, then designing and carrying out experiments and analyzing results; but rather by a messy and nonlinear process of tinkering with experimental systems, then iteratively designing, testing, and modifying those systems. The design of learning technologies is more often composed of iterative cycles of designing, modifying, and testing the technologies in parallel with a series of problems to which the technologies relate.

In the case of mobile learning games that we have developed, on the technology strand we have examples that are firmly grounded in connecting people with each other or with their surroundings or both as core components of the games. This design approach arose from early experience with mobile games, informed by theories of learning. In parallel, we have goals of creating experiences that promote the new set of skills demanded by the twenty-first century while meeting the realistic constraints of classrooms, schools, and other learning environments.

What ties the goals (twenty-first-century learning) and tools (mobile games) together are two underlying learning theories that guide much of our work—*constructivism* and *situated learning*.

*Constructivist theory* (Piaget 1977; Bruner 1986; von Glasersfeld 1995) states that people learn by constructing their own understanding of principles and phenomena. They build that understanding based upon past experiences and beliefs, integrated with current experiences. That is, according to constructivist theory, learning is an iterative process of updating existing understanding with new information acquired through activity. Educational activities informed by constructivist theory recognize that learners enter these activities with preexisting knowledge, and shape that knowledge through the experiences of those activities. Thus constructivist activities are characterized by wide open spaces to explore, room for learning through both success and failure, feedback that learners can use to adjust their own understanding, and multiple possible outcomes. Constructivist activities often take the form of problems that learners are motivated to solve in unique and active ways.

*Situated learning* (Lave and Wenger 1991; Lave and Chaiklin 1993; Wenger 1998) is a theory that describes the process of learning as highly social, embedded in the lives of learners, and can be complementary to constructivism. Much of the theory of situated learning centers on the notion of *communities of practice*: dynamic groups that are present throughout our lives in which we participate in various ways. Such groups exist in schools, workplaces, social organizations, and families. Each of these groups has a set of practices that members learn over time. While this training may sometimes be formalized (such as joining an outdoor group to learn rock climbing), it is often much more tacit (such as spending time with a group of friends and learning what they enjoy). In addition to the set of skills and knowledge one might learn by participating in such a community, the cultural and social practices are also part of what is learned.

Learning activities that draw on the principles of situated learning and communities of practice share many characteristics with constructivist activities. They are also often problem-based, and draw upon previous experience in the learner's life. These problems usually are inherently meaningful and motivating to the people involved. But situated learning also explicitly draws upon the real-world context in which the problems are set and the community that is either developed or appropriated around the activity. Social activity and connections to the real, physical world are important characteristics of situated learning.

The synthesis of the constructivist and situated learning paradigms lead us to design activities that are inherently social, authentic and meaningful, connected to the real world, open-ended so they contain multiple pathways, intrinsically motivating, and filled with feedback. While many technologies can foster some of these design elements, mobile learning games are particularly well suited to supporting them all:

Social Mobile games draw upon existing social relationships and means of communication when situated in social environments that participants already know how to negotiate. Many technologies require developing new means of communication and of fostering social relationships. Mobile games do not need to reinvent these systems, as they draw upon preexisting ones—whether those be face-to-face, through other modern means like instant messaging, or even phone calls. This real-time connectivity can increase the building of real social relationships among group members.

Authentic and Meaningful Situating mobile games in real contexts connects them to actual people, places, and events. While the specifics of these games may be fictionalized, basing them in the reality of physical places—through which the players must physically navigate—deeply connects the players to the problem and place at hand.

Connected to the Real World   While PC and console video games strive for high-fidelity graphics and game-play experiences, mobile games that are already set in the real world get 100 percent real-world fidelity for free. Many intangibles of the physical world get incorporated in mobile games without being explicitly designed into the games.

Open-Ended/Contain Multiple Pathways   Like real-world problems, the problems designed into my group's mobile games are significant and characterized by the lack of one clear answer. Navigating large geographic and information spaces makes knowing all of the information impossible, and players must constantly redefine their own goals—in the end defending their answers and the means that they used to define that answer.

Intrinsically Motivating   Mobile gaming not only is the fastest growing gaming sector, but also has the broadest appeal across genders, ages, and interests. Drawing upon some of the design principles that have brought about the broad success of this platform, we can make learning games that are equally broad in their appeal.

Filled with Feedback   Feedback in mobile games can come in many forms. The most obvious is directly by electronic means. As players physically move around in the real world and interact they are provided with virtual feedback based on their actions, preprogrammed outcomes, and underlying models. But feedback can also come from other players. As players encounter each other, by design and serendipity, they exchange information and ideas, providing useful feedback to each other. Finally, new means of feedback can be discovered by players as they appropriate information from their physical reality with the information provided to them by the games.

These characteristics fit well with the twenty-first-century skills we are trying to build because they provide the appropriate context for students to develop these complex abilities. Thus mobile learning games may be uniquely poised to address these skills in schools, in the workplace, and in life.

## Who Should Read This Book?

The focus of this book is on the research and design of mobile learning games. In order to effectively design such games, they need to be set properly in the context of schools and learning, games and education, and mobile games. The first half of the book lays out the history and current state of each of these areas. Chapter 1 is about schools and education reform, and how technology fits into the academic world. Chapters 2 and 3 are about educational games, both historically and by design. Chapters 4 and 5 are about mobile games in general and applied to education in particular. The remaining chapters (6 to 12) offer case studies in the design, implementation, and research of mobile learning games.

I'd like to think that the book can be read through the lens of anyone interested in schools, learning, educational games, or mobile games, and that there should be lessons and stories pertinent to any and all of these realms. Even the case studies that straddle intersections of these topics can be teased apart to provide insights for people with interests in one or more fields. Thus teachers, trainers, technologists, administrators, instructional designers, educational and game studies researchers, and game designers should all find stories of interest to them. For teachers and other educators, this book provides a new perspective on how educational technologies can make a difference in (and out) of the classroom. Even if mobile learning games are not specifically feasible or desirable, it is useful to understand the salient characteristics and practices that make computer technologies powerful and feasible classroom learning tools. For trainers who provide professional development and just-in-time or on-demand learning for the workplace, this book introduces new tools and approaches that should be considered when looking to enhance the knowledge, skills, and performance of a diversity of workplace learners. Training adults to enhance their job-related expertise (everything from first responders dealing with a crisis to managers working with teams) is a challenging task, and these technologies may prove effective in reaching this audience. Educational technologists can similarly benefit from understanding what makes mobile learning games effective tools, and using these principles when evaluating new learning technologies. Administrators may look to the lessons of implementing mobile learning games when planning their own technology infrastructure, both for what should be in their plans, as well as what not to invest in. Beyond technologies, the material covered in this book also speaks to the design of classroom and learning activities more generally, from which instructional designers could benefit. Parents can begin to understand more deeply what their kids are doing and learning when playing mobile games, and gain some insights that might help guide decision making when it comes to choosing video games. Researchers in educational technologies and game studies can get a primer in a segment of the video game space that has received little attention from either field, and see the possibilities for research and development. And game designers may look at their work differently and push into a relatively untapped portion of the market, as they better understand the growth potential and interesting design challenges and opportunities of mobile games.

# Acknowledgments

Throughout this book when I mention the work that *we* have done, it is not use of the "royal" we, but rather refers to the work that students and staff of the MIT Teacher Education Program (TEP) have contributed to, as well as the work that I have conducted with colleagues. The MIT TEP prepares MIT students for careers in education and designs, develops, and researches educational technologies. It is also a partner in the Education Arcade. The case studies that I describe are truly collaborative efforts, and the work of everyone who participated should be recognized. I'm not quite sure I can name everyone, but I'll try.

At the base of all of the work that we have done are countless hours of software development by MIT undergraduates (known as UROPs on campus). I have been fortunate enough to work with quite a number of talented and dedicated UROPs over the years (on the game projects noted in parentheses), including Kodjo Hesse (Environmental Detectives), Philip Guo (Participatory Simulations), Ben Povlich (Augmented Reality), Victor Costan (Participatory Simulations and Palmagotchi), Marios Assiotis (Mystery @ the Museum), Nicholas Behrens (Outbreak @ the Institute), Jon Hyler (Mystery @ the Museum), Dustin Bennett (Augmented Reality), Mark Boudreau (Augmented Reality), Paul Soto (Outbreak @ the Institute), Robert Kwok (Outbreak @ the Institute), Hector Beltran (Outbreak @ the Institute), Kirupa Chinnathambi (Augmented Reality), Xia Liu (POSIT), Tze Kwang (POSIT), Neil Dowgun (POSIT), Mahmood Ali (POSIT), Mark Vayngrib (POSIT), and Robert Reyes (POSIT). The master's degree students, who contributed at least a year of their lives to the cause, include Priscilla Cheung (Environmental Detectives), Tiffany Wang (Augmented Reality), and especially Ben Schmeckpeper who really helped the Augmented Reality project turn the corner due to his year and a half of work on the project.

The staff of the Teacher Education Program have guided these projects through thick and thin, providing game design, project management, research support, and a great deal of well-placed cautious optimism. Jen Groff worked on the Participatory Simulations project, and also provided valuable research support on this book. Eric Rosenbaum worked on both Outbreak @ the Institute and POSIT, and filled in just about everywhere else. He is truly a jack of all trades (and master of many). The queen bee of the lab (and also the first TEP employee), Judy Perry, has played a significant role in each and every one of the projects mentioned here, and has provided guidance, humor, and a lot of energy to make all of this work a success. This core group was supported by others over time, notably Louisa Rosenheck and Britt Boughner on the POSIT project.

A number of colleagues have helped support this work both directly and indirectly. Kurt Squire at the University of Wisconsin deserves half the credit (or more) for the Augmented Reality work. Kurt came to me a number of years ago with a desire to do something interesting in the mobile educational games space. Many hours of brainstorming and discussion eventually led to this idea, and he has since continued to be an invaluable collaborator and colleague. Henry Jenkins and Alex Chisholm were also a part of the Games to Teach project that helped spawn this platform and have remained great supporters. Philip Tan here at MIT has been brilliant in coming with ideas for game mechanics. Chris Dede at Harvard University has become a great champion of all learning games mobile, and is an insightful and innovative partner. Matt Dunleavy, who has worked in Chris's group, continues to make great contributions to the augmented reality cause, taking this platform to new people and places. Susan Yoon, my former postdoc, now at the University of Pennsylvania, did all of the foundational work on opinion dynamics, upon which several of my projects are based. I thank her not only for providing that intellectual basis, but also for the rigor that she has brought directly and indirectly to many related projects (and also for acknowledging that there are no truly neutral opinions). Ginger Richardson and the Santa Fe Institute have provided a great intellectual space for exploring complex scientific ideas, many of which spawned elements of the games mentioned here. The original participatory simulations emerged from the work of Mitchel Resnick's student Vanessa Colella at the MIT Media Lab. Mitchel continues to be a superb colleague, and deserves credit for supporting my career advancement at MIT through this body of work. On a similar note, my department (The Department of Urban Studies and Planning) deserves thanks for acknowledging the value of this research and enabling me to continue this work. Megan Murray at the Harvard School of Public Health deserves thanks both for her help on epidemiological models, and for allowing us to work with her epidemiology classes. Jon Guryan at the University of Chicago and Brian Jacob at Harvard lent interdisciplinary expertise to help design and frame some of the Participatory Simulations research.

Of course the work of all of the aforementioned people needed to be supported, and for that I have to thank Microsoft Research, The Intel Foundation, The U.S. Department of Education, an anonymous donor (who came through Dick Light and Tom Healey of the Young Faculty Leaders Forum), and the Scheller family, which has provided ongoing support for the Teacher Education Program.

My editor, Douglas Sery at the MIT Press, made this whole process quick and enjoyable.

Finally I have special family members to thank. Several of them I thank for the anecdotes they provided that are included in various places in this book. My parents, both career educators, have taught me what teaching is. They also served as a model demographic for this book, being "teach-savvy" but not particularly "tech-savvy." My in-laws provided a model of another sort, lending their experience in how an academic family gets through the writing of a book. They also lent a hand a number of times helping with the kids so I could write. And (this time I will not forget) I have to thank my wife Rachel for her many years of encouragement, which enabled me both to find and pursue my academic passions. She gave me the courage to take on this big project, and may now finally forgive me for inadvertently not thanking her in my Ph.D. thesis over a decade ago.

# Augmented Learning

# 1 Educational Innovation through Time

I peer through a portal into three biology classrooms—one today, one twenty-five years ago, and one fifty years ago. As I watch the first class I see a group of students pour into the classroom and settle into their wooden seats assembled in rows along the lab tables. They sit down, take out their notebooks, and prepare to copy notes from the teacher. The teacher walks into the classroom, takes out her notes, and begins to lecture. She writes notes on the board, and illustrates concepts with diagrams, also on the chalkboard. If it weren't for the hairstyles and clothing, I might need to watch for quite some time to determine which class is which. The instructional style has hardly changed in the last fifty years. If I happened to watch on a day in which the class had a test, I wouldn't have any greater luck in determining the era of the classes (aside from the quality of the reproduction of the test on photocopy) as assessment methodologies haven't changed much either.

Of course there are superficial differences. When showing illustrations, the class from fifty years ago uses only white chalk, the class from twenty-five years ago uses multicolor chalk and overhead transparencies, and today's class occasionally uses illustrations from CDs projected via LCD projector.

The striking thing is that there are tremendous differences in content. Fifty years ago the recent discoveries of Watson and Crick about the structure of DNA had not yet made their way to the classroom. Without DNA, Mendelian genetics were a bit of a black box and taught as such. Twenty-five years later DNA was something that was observed and well understood. The connection to genetics was taught, as was the mechanics of how DNA works. Jump forward another twenty-five years to the present and the number of classes about DNA has increased ten-fold and they now include information not only on how DNA works but also how it is manipulated and applied in science and medicine. The technologies (at least as they are presented in a book) are taught and tested in detail.

The relevance of the science has also changed dramatically over that fifty-year period. Fifty years ago biology was something that only doctors needed to learn into order to practice medicine (and something that physicists needed to take in order to fulfill requirements). Twenty-five years ago biology was still the poor relative of physics, but was a promising field of the future for scientists. Today biology and biotechnology affect not only scientists and doctors, but also citizens every day as they must weigh in on issues like stem cell research and assess their risks in the face of emerging diseases such as bird flu.

Ask any teenager—sitting in front of a computer—what DNA replication is and within thirty seconds with the help of Google or Wikipedia they can regurgitate a description that would pass most classroom tests today. Within the first few hits on Google you'll find definitions, animations, and activities that explain the process better than it was explained to me in high school or university.

Students today could similarly look up information on avian influenza, commonly called bird flu, and tell you about the genetic structure of H5N1 (the scientific designation for the strain of bird flu). But ask them to assess the risk of them getting infected by bird flu, and they are not likely to even know where to start. Online they can find many estimates by different organizations over time. How do they make sense of that disparate set of sometimes conflicting data? How do they break this question down into the necessary components to evaluate? For example, they could get bird flu from a bird, or another person. How would that happen? What is necessary for that to take place? And what is the biology behind these varied pathways? And then they must consider how they could estimate their individual chance of infection. How do they assess risk and use infinitesimally small probabilities? What do those numbers mean, and how do they pertain to individuals? How do they understand the different ways in which a person might come into contact with this disease? Can they examine their own social networks to understand the inherent heterogeneity across geography and subpopulations? What about vaccines and cross-immunity? Are all people equally susceptible and likely to come into contact with this virus?

The student in the biology class is hardly any more prepared to tackle these issues and provide an estimate than the student who has merely looked up a few facts on the Internet. And both of these students are only marginally more prepared to answer this question than their peers viewed through the portal twenty-five or fifty years ago. Yes, there are some concepts and terminology that the students from today may be able to recall, but that is the easy part of this problem. The hard part is breaking this into the component questions, understanding how these relate to each other, managing to make decisions with incomplete information, knowing which experts and sources to draw upon and trust, understanding systems and networks, analyzing data, interpreting numbers and graphs, and choosing and using the appropriate information technologies to make this possible.

Those are the hard tasks. Most twenty-first-century schools do no better a job at providing these skills than do the schools of twenty-five or fifty years ago (Hirsch 1996; Ravitch 2000). Yet the economy and the world itself are qualitatively different than they were in those eras. We currently face many new challenges in our educational system, yet maintain similar dilemmas from those previous times.

## U.S. Education Reform

The schools through the portal all exist at critical points in U.S. educational history. Fifty years ago schools were refocusing their efforts on science and math education during the Sputnik era. There was a national imperative on American competitiveness during that time. The United States needed to create a workforce that was capable of competing with the Soviets. While schools have traditionally been under state and local control, the federal government stepped in to boost efforts in schools to create a workforce with the knowledge deemed necessary to stay competitive. This marked a significant milestone in stressing particular content in the classroom.

By many metrics this educational initiative met with great success (not the least of which was putting a man on the moon first), producing the "baby boomer" generation that went on to produce great leaders in industry and government and in many ways defined the United States in the latter half of the twentieth century. Yet the schools of this time, as throughout much of American history, faced challenges as they tried to weigh competing priorities and conflicting philosophies. The mission of public secondary schools grew at this time to include accommodations for *all* students, not just an elite few. It also balanced the goal of training the future workforce with producing good citizens and creative thinkers.

The students in the room from twenty-five years ago are living in a different educational landscape, though one that faces many of the same challenges. Their classroom may be one of those harshly criticized in *A Nation at Risk* (National Comission on Excellence in Education 1983), a comprehensive report by the United States government published in 1983. This report was prepared by a panel of educational and industry experts to assess the current state of schools in the country at that time. There was the sense that schools were not adequately preparing students, and this report was an attempt to determine the extent of the problem. The fear was that the United States was falling behind Japan, South Korea, and Germany in key industries, and that the nation's schools were likely contributing to the problem (or could be a key to turning the situation around). *A Nation at Risk*, as evidenced by the title, in fact found that the schools, teachers, and students at that time were underperforming. They were scoring low on international benchmarks and standardized tests within the United States. Illiteracy was high and on the rise in both students and adults. Competencies in "higher order" thinking skills were very low. From the industry side there were complaints that employers consistently were required to train their incoming workforce with remedial education. Leaders expressed alarm at the time, particularly in light of the growing importance of science and technology in the workplace and lives of citizens, and the glaring deficiencies in understanding these rapidly changing fields in all but an elite minority. A series of recommendations were made through this study, which included a focus on a broad curriculum (notably including adding to English, social studies, science, and math the fifth discipline of computer science), new teaching methodologies, and metrics for success.

While many outcomes may be associated with this report, the standards movement is the most prominent outgrowth of both this report and the rationale for developing it in the first place. Greater standards needed to be put in place to ensure that students would meet minimum competencies in schools, and tests needed to be created to measure if those standards were being met. Great emphasis was placed on ensuring that all students had the opportunity to meet these minimum competencies. The culmination of the standards movement was the creation and enactment of the No

Child Left Behind (NCLB) Act of 2001, which placed the responsibility of setting and measuring standards in the hands of each state (without additional funding as the states are quick to point out). Those standards, however, had to meet federal guidelines in order for states to continue to receive federal funding.

Today those standards are firmly in place, yet there is still a perception that the country is again slipping behind. The nations we now fear are China and India. These massive countries, and their associated economies, have created the perception of an economic threat and have therefore raised a new national imperative to enhance the competitiveness of all citizens. As made well-known by Thomas Friedman in *The World Is Flat* (2005), many of the industrial and manufacturing jobs have moved overseas or been displaced by automation. Further, many of the clerical, support, and technical positions that had been a staple of the U.S. economy have also moved overseas. The result of this shift is that the schools that were preparing students for a world in which they could depend on those jobs are obsolete and must adapt.

## Modern Skills and Modern Education

A collection of some of the nation's leaders in education, industry, and government formed The New Commission on the Skills of the American Workforce. In 2006 this commission released the report *Tough Choices or Tough Times* (National Center on Education and the Economy 2007), which provided a harsh criticism of the current state of schools in the United States, and how they were failing to prepare students for the modern global economy. The commission argued that the skills that the standards movement has pushed are no longer sufficient to remain globally competitive. Those skills must be bolstered by a new set of skills:

Strong skills in English, mathematics, technology and science, as well as literature, history, and the arts will be essential for many; beyond this, candidates will have to be comfortable with idea and abstractions, good at both analysis and synthesis, creative and innovative, self-disciplined and well organized, able to learn very quickly and work well as a member of a team and have the flexibility to adapt quickly to frequent changes in the labor market as the shifts in the economy become faster and more dramatic.... The core problem is that our education and training systems were built for another era, an era in which most workers needed only a rudimentary education. (Executive summary, *Tough Choices or Tough Times*, p. 8)

Others have similarly advocated that today's workplace demands markedly different skills than students currently are being prepared for in schools (see for example Murnane and Levy 1996; Levy and Murnane 2004). While schools do a reasonable job of preparing students in the "hard skills" (math, literacy, geography), they are not adequately preparing them in the "soft skills" (problem solving, communication, working in groups, and so on). Jobs increasingly are relying on these soft skills, as the hard skills alone are insufficient for managing the diverse tasks in the modern workplace.

Levy and Murnane (2004) identify two particular sets of skills that the modern workplace relies heavily upon—expert thinking and complex communication. *Expert thinking* applies to domain-specific problems that cannot be solved by following a simple set of rules. Complex problems often cannot be solved in a linear fashion, and require deep understanding of systems and processes, an understanding that comes with intricate familiarity with a particular area. This kind of thinking is not easily taught in the format of current schooling that emphasizes following rules, and superficial coverage of vast amounts of content. *Complex communication* describes a variety of personal communication skills necessary in many aspects of work. These includes persuasive speaking, inductive reasoning, communicating with colleagues, understanding clients, making inferences, and describing technical work in nontechnical ways. Again, students lack the opportunity and instruction in contemporary classrooms to hone these skills.

There is a strong overlap between these skills and information technologies. Much of expert thinking and complex communication relies on information technologies. One may assume that this is one place that current schools excel. Looking around at students in or just out of high school seems to show strong evidence for their abilities to use information technologies. And if we were simply to measure the ability to employ software and communication technologies to complete simple tasks, then the students would be doing okay. But this set of "contemporary skills," as defined as a part of Fluency with Information Technologies (FITness) by the 1999 National Research Council study on *Being Fluent with Information Technology* (and revisited in 2006 in NRC's *ICT Fluency and High Schools* [National Research Council of the National Academies 2006]), is just one component of preparation for using information technologies. In addition to contemporary skills are "fundamental concepts," the basic understanding of the workings of technologies, and "intellectual capabilities," a suite of cognitive abilities that must be learned in order to know how to apply information technologies to relevant tasks and problems.

Learning all of these components of information technology competencies (FITness) is critical to being prepared for the modern world, both in the workplace and out. This set of intellectual capabilities comprises a broad range of skills related to living and working in an IT-inundated world. None of them specifically mention technology, but all are tightly integrated with it. The skills (National Research Council 1999) are as follows:

1. Engage in sustained reasoning.

2. Manage complexity.

3. Test a solution.

4. Manage problems in faulty solutions.

5. Organize and navigate information structures and evaluate information.

6. Collaborate.

7. Communicate to other audiences.

8. Expect the unexpected.

9. Anticipate changing technologies.

10. Think about information technology abstractly.

This list looks distinctly different than the frameworks and standards set forth by most states to establish guidelines for what students need to know. While those sets of standards often pay some attention to scientific thinking, critical thinking, data analysis, and communication skills, these are most often the skills that are ignored on the assessment tools, and subsequently in the classroom. Instead, lists of facts and domain elements are ticked off as lessons touch on those concepts. In the rare cases where technology enters the assessments it is in superficial ways, asking students to define particular technologies or other tasks that are in the IT concepts category, squarely missing the suite of intellectual capabilities that will be required outside of school (in the workplace, in post-secondary education, or in being informed citizens).

In revisiting these skills (National Research Council 2006), experts have attested to the continued importance of these intellectual skills. While the particular technologies and concepts have changed (in 1999 the Internet was still in its infancy, without blogs, streaming video, or social networking), the intellectual skills are still critically important and are likely to remain so for a long time. However, further looking at this list makes apparent the difficulty in teaching these skills in the current school modality. With vast amounts of content to cover, how does one have the time to "engage in sustained reasoning"? How does one go deep enough into a single topic to learn to "manage complexity" or "manage problems in faulty solutions"?

Students cannot simply be taught FITness. You can't just provide an extra week of instruction on these topics, or even an extra class on this set of skills. Building FITness cannot be done abstract and discretely, but requires integration with existing subject matter. We need to find ways to provide students with meaningful experiences through which they can develop these skills in the context of their existing subject matter and coursework. The skills must also be learned in pursuit of improving skills in expert thinking and complex communication.

## IT as the Problem and the Solution

As much as IT has created a complex intellectual landscape that can only be navigated with a new set of skills, it has also provided the means to learn that navigation process, and at the same time make advances in the educational system that have been slow to come. Using new technologies we can engage students in deep, meaningful, realistic, and relevant problems, the kinds of complex collaborative problems that education reformers have been clamoring for for many years. Some of this can change through the use of desktop/laptop computers. Students can now work with data analysis tools, collaborative learning environments, simulations, multimedia authoring tools, and virtual environments, all of which offer access to new content and new ways of learning. Video games (see chapters 2 and 3) are particularly promising technologies for learning. This young field is just beginning to explore how games not only can motivate students, but also provide rich learning environments that challenge students in understanding complex problems, whether by jumping back in time to understand history, or inside of a blood vessel to explore the human body. Gaming technologies have a lot to offer despite their previously tenuous relationship with learning.

While some change can come through the use of traditional desktop computers in computer rooms, this setup in and of itself has become problematic. Rather than using tools such as the ones mentioned above to teach a new set of skills, most computer rooms have been appropriated to teach all of the old skills in slightly new ways (see chapter 6). Students collect information for reports from the Internet instead of the library, they type reports on word processors instead of typewriters, and make multimedia presentations on PowerPoint instead of poster board. But the content and skills are largely the same. They are often not provided with the opportunity to explore emerging ideas, work in distributed teams, take on complex issues, or take ownership of problems at hand. All too often students still are not engaged in their own learning. A famous study (Wenglinsky 1996) investigated the connection between different uses of computers in the classroom and academic achievement as measured by performance on the fourth- and eighth-grade National Assessment of Educational Progress (NAEP). Wenglinsky found that students who used drill and practice software for rote tasks did the same or worse than students who didn't use computers at all. However, math and learning games along with simulations and applications were associated with higher scores for students, with the added benefits of increased attendance and higher school morale. But we need to find ways to promote that kind of learning without getting bogged down in the routine of computer rooms.

What this means is that we need to harness IT for learning *outside* of the computer room. We need to put technology into the hands of teachers in their own classrooms, where they can use it to enhance and integrate with their own instruction, not simply add an IT component to their existing courses. Handheld computers, or personal digital assistants (PDAs), like the Pocket PC/Windows Mobile or Palm platforms provide such an opportunity. Relatively inexpensive handhelds can provide a one-to-one (one device per learner) solution in the classroom now. While some researchers and software designers have sought to make these serve the purpose of traditional computing resources, only in a smaller form (see chapter 4), the real opportunity comes in integrating computing with other social, collaborative learning activities that allow technology to facilitate learning and problem solving, rather than being the focus of such efforts. One successful application of handhelds in education of this vein has been the use of Probeware, which allows students to take their handhelds out of the classroom and into the field where they can sample data about the local environments. However, the rise of mobile gaming on cell phones, Game Boys (handheld game consoles), and other mobile devices (chapter 4) suggests that handheld computers can play an important role by bringing gaming technologies into the classroom (chapter 5). The handheld platform also avoids the stigma too often associated with video gaming technologies, by decreasing isolation of students from each other, and facilitating new gaming patterns and genres.

### Building FITness through Handheld Games

For a number of years I have been designing mobile games for learning that push the envelope of both what students learn "in" school and how they learn it. Much of this technology has subsequently been employed out of school in universities, in museums and other informal learning centers, and increasingly with adult learners. These games have been designed with FITness skills as a key set of desired learning outcomes. While the initial chapters of this book set the stage for the role of handheld games in learning, the later chapters explore the design and implementation of these games for a variety of learners. I set out to highlight key lessons learned through both the design and subsequent study of these games, and distill some design principles that have been particularly effective at targeting the FITness skills. I review these skills now in more detail, noting how they map to mobile games and where they are discussed further in the chapters to follow:

1. Engage in sustained reasoning. Learners engage in problems taking many days to solve, and must employ a variety of resources for problem solving. This is ture of most of the games that have been developed; see for example the participatory simulations of chapter 6.

2. Manage complexity. Learners must try to manage complex systems with unpredictable outcomes, such as epidemics (chapters 6 and 9) and virtual ecologies (chapter 12).

3. Test a solution. Learners encounter problems for which they must not only assess the situation but also design (see for example chapter 7) and implement (chapter 10) those solutions.

4. Manage problems in faulty solutions. Learners must understand that their interventions will not always lead to the perfect result. Sometimes

trade-offs must be made in designing solutions (chapters 7 and 8), guided by the feedback learners receive in implementing solutions (chapter 10).

5. Organize and navigate information structures and evaluate information. Learners must collect and evaluate information from a variety of sources, including primary data, documents, and witnesses, as demonstrated in many of the location-based augmented reality simulations. This task is particularly relevant to the games described in chapter 11.

6. Collaborate.   Collaboration is indeed one of the key design principles that cuts across all of the games that we have designed. No game can be played alone. Chapter 9 includes a thorough discussion of this design principle.

7. Communicate to other audiences.   Learners must be able to communicate with other players within the games (role playing different audiences as in chapter 11) and often have to present their findings to panels as part of the "end game" (as in chapters 7 and 8).

8. Expect the unexpected.   Planning for the unexpected may seem like a paradox, but in managing complex systems, and understanding how they work, learners can begin to assess risk and plan for such contingencies as they must do in the games described in chapters 10 and 12).

9. Anticipate changing technologies.   While the technology is not an explicit focus of any of the games that we design, learners must master a range of technologies including peer-to-peer messaging, wireless communication, spatial navigation, data collection, and analysis and visualization. The diverse landscape of changing technologies is represented across all of our work.

10. Think about information technology abstractly.   Again the focus is not on the technologies themselves but how they are used. Thus we don't teach about the technology, but rather help learners understand the functionality that the technologies provide by presenting them with a problem and the means to solve that problem.

The themes of *expert thinking* and *complex communication* are also present in much of the work that we have conducted. Using role playing, learners in our games examine problems through the eyes of professionals in many fields, developing the expert thinking required to solve these problems. Similarly players must communicate with each other (often representing different roles and skills) and to outside audiences through a variety of media. They learn about new technical, social, scientific, and political concepts in order to understand and convey these messages to other players through complex communication.

## The Trojan Mouse

The next few chapters outline in more detail the case for handheld games for learning. They explain the sordid past of games in education (chapter 2), what games really are (chapter 3), the history of handheld games (chapter 4), and ultimately the small world of educational handheld games (chapter 5). Those chapters are followed by narratives that describe our work in designing, developing, and studying handheld games for learning.

But before proceeding into those chapters I'd like to contextualize my arguments for the use of handheld games in learning with these observations:

Handheld games are not a panacea.   While I believe that handheld games afford great potential for learning, and are particularly well suited to address some of the issues previously outlined in a way that is both compatible with current schools and capable of changing them, handhelds are not the silver bullet to save education. They can, however, play an important role.

Handhelds can play nicely with other technologies.   There are many technologies that can and should have a place in the educational arsenal of learning organizations. Collaborative tools (blogs, wikis, knowledge-building software), immersive environments (e.g., virtual worlds like Second Life), media processing and sharing tools, and many others have a home in education. Handheld games fill a new and less-well-known niche in this ecology of tools.

It isn't all about the technology.   Most of the intellectual capabilities previously defined are relevant to understanding most modern issues and problems. They need not necessarily be associated with technology at all. Many of these skills are equally relevant to constructivist learning that has been promoted by education reformers for decades, and could be fostered without technology. Technology, however, is the vehicle for getting these intellectual capabilities into schools discretely. Others have called this use of technology the "Trojan Mouse." Though that metaphor doesn't work as well on handhelds, which employ a stylus, not a mouse. The "Trojan Stylus" just doesn't have the right ring to it.

# 2 Educational? Games?

It would be difficult or impossible to write a book about educational games that didn't ask and attempt to answer: What is "educational"? And what is a "game"? The latter question, of what defines a game, has been taken on by a number of games scholars in the last several years (e.g., Salen and Zimmerman 2003; Juul 2003; see chapter 10 for further discussion). Such definitions often contain some combination of rules, goals, outcomes, and, sometimes, fun.

In fact, an activity that I do annually in my own class on educational games is to have students attempt to define "game." We start by brain-dumping recollections of every game that everyone has played without attempting to define what a game is. What results is a compiled list containing hundreds upon hundreds of games, from tic-tac-toe to World of Warcraft to beer pong to games that may best be classified as mind games (like "Can I get that girl/boy to like me?"). Despite the controversy that this question has created in the scholarly community, in a class of a dozen or more students there is little debate on any of the games that students include in their lists (with the possible exception of the mind games mentioned above). When we move from definition by example to formalize the definition, the debate begins.

There is some debate on rules, especially considering in the mix a game like Nomic, which has an evolving rule set. Nomic (Hofstadter 1982), in case you are unfamiliar with it, is a game in which the rules involve changing the rules. Each round a player's move is to make a rule change. This has spawned a large genre of such rule-changing games. But regardless of rule changes, there are rules. More debate centers on issues of "fun" and "goals." Goals become a challenge when considering gaming spaces like Second Life or even The Sims, in which goals often are not specified by the game itself, which instead provides feedback that the players can use to establish their own goals. In terms of fun, perhaps just good games are "fun." Or maybe "fun" is in the eye of the beholder.

## Defining "Games" in Schools

It is actually this flexible definition of game that becomes an asset when we're working with games in the classroom. For all practicality, when dealing with students, teachers, and learners, we can use a functional definition of game and say it is anything that the participants will let us get away with calling a game. More precisely I can define it as a "purposeful, goal-oriented, rule-based activity that the players perceive as fun." In the context of school, where fun is not typically a priority design feature of daily activities, this definition provides a great deal of latitude. In fact, in many classrooms the only experience that is typically passed off as a game is hard to distinguish from a test. Students divide up into two or more teams and answer questions (often of the type that they'll be seeing on a test or quiz a few days later) from the teacher, such as "What is the organelle responsible for photosynthesis in plants?" Sometimes, to make the experience even more game-like, the teacher will provide the answer and the students must respond Jeopardy style: "This is the organelle responsible for photosynthesis in plants." The sense in which this is a game instead of a test is that there are points awarded for each question, and the outcome is a prize instead of a grade.

I don't want to trivialize the teaching profession (I was a teacher and know how hard the work is), or to overstate the ease with which one can create more engaging games in the classroom. There are many constraints on the daily activities of classrooms, including the demands from standards mentioned in the last chapter, and the ever-present demands of 150 or more students on a daily basis. Most teachers would like to have tools through which they could easily motivate and educate students, but there is a paucity of such tools.

Teachers know that the word "game" has a lot of appeal and they try to motivate students through games such as the one I described, or even worksheets or other activities that are only games in that the teacher calls them "games," hoping to motivate the class to engage in them (e.g.,

"Class, today we're going to play a game. Put away your books and write your name at the top of the test, er, game."). However, that same term also can be very threatening. In a time when the words "video game" are most often mentioned in the media in an unflattering context, such as for allegedly inciting violence, portraying pornographic images, or causing addiction, some teachers fear that calling something a game that actually might *be* a game is undesirable. We are often asked by teachers to refer to our projects as "simulations" or "activities" instead of "games." A department chair at one of the schools we worked with recently sent this request before our meeting with the rest of the teachers in that department: "Is there any possibility that we could present this as something other than a 'game'? I am just concerned about the message that will send and would be thrilled if maybe we could categorize this as an investigation, debate, exploration, etc. I think it would be more palatable in the context of the work we are engaged in at [our school—name removed]."

This response has an interesting duality. On one hand it is quite critical of games, suggesting they will be disruptive in light of the other positive work that the school is engaged in (and there is a lot of it). One may assume that either the games are expected to cause the students to see their work in a less serious manner, which would compromise their investment in learning, or that employing games would signal to teachers that we are not taking their work seriously. So it may be the case that teachers value what games actually provide but discount the term as trivial. On the other hand, when teachers suggest alternative terms, the terms they suggest convey rigor and deep learning associated with games. The terms "investigation," "debate," and "exploration" all have meaningful and valued associations in school. Teachers often make great efforts to engage students in these kinds of activities, and the use of these terms as alternatives is a testament to the implied value of this medium.

In another school we worked with a teacher who was using a number of our activities in his science class. There were multiple sections of the same course, and a different teacher was teaching a traditional version of the same class in another section. The students in the traditional class complained to the principal that the students in the other class were "playing games" while they were doing book reports, and that it wasn't fair. The principal snapped into action immediately and told the teacher we were working with to stop his activities and have the students do more book reports.

In these cases we usually try to do a little intervention and offer some media education about what games are (and are not). But in the end, I'll usually bend and play down the use of the word game if it is requested and call them handheld-based activities or simulations. But when we get in front of the students, the word game goes much further. They can adapt what they already know about "playing" games and start plotting out strategies, incorporating what they know about collaboration and competition, and define their own goals with little instruction. They readily buy into the concept that these are games because they are indeed fun (by their own words and measures). Despite this label, I have yet to see these games cause any disruption of the kind that teachers fear, either with students or other teachers, which might interfere with the work in which they are engaged. To the contrary, we find that use of the label often motivates the unmotivated student, and intrigues the motivated student in a new way. These experiences are often cited in end-of-year evaluations as having been memorable and motivational. In short, they can be both "games" and "productive" at the same time.

### Are All Games Educational?

But it is certainly a challenge for designers and creators of educational games, as well as many of the teachers who are their potential consumers, that the game label can be a stigma as much as it is an asset. One doesn't need to look far to see where that negative view is coming from. The media is filled with coverage portraying video games as the source of malevolent content and behavior. This stereotypical portrayal of video games has become so deep-seated that many people don't even know they hold this view. My family and I were recently visiting friends who had an elementary school-aged daughter, who I'll call Neera. I saw that her parents had given Neera a computer, and it was placed at a low, small table to make it very accessible to her. Neera was eager to show my young son what she could do on her computer, and my son was quite happy to oblige (as he has enjoyed computer games since he was three). They played a Disney Princess adventure, or something like that, which they both enjoyed. In my quest for good children's software (examples of which are few and far between), I asked one of Neera's parents what *other* games they had. The parent replied, "Games? We don't let her play video games. No violence." As with the previously mentioned teachers, I dropped the term and used some other term such as computer "exploration" or "investigation."

What was clear was that to this parent, video games involved shooting things. A Disney Princess adventure therefore was not a video game. This is a problem that the whole video game sector faces, not just the small educational sector. The message about the value of games and the vast ecology of different games need to be disseminated in order for games to have a chance in the classroom, or in educational venues more broadly.

That raises the question, What is "educational" about games? Others have argued (e.g., Prensky 2001; Gee 2003; Squire and Jenkins 2003; Johnson 2005) that there are many aspects of commercial off-the-shelf (COTS) games that model the way that good learning happens. Even playing "violent video games" (in the words of Gee—though he is not necessarily advocating for violent video games) models how we should structure problem solving for students—by challenging them to transfer knowledge from puzzle to puzzle, learn about systems, test hypotheses, and to communicate effectively. For example, in trying to kill a "boss" (one of the big, nasty characters that usually appears at the end of a level in "first-person shooter" games) in an upper level of a game, a player might need to transfer some of the lessons that they learned in earlier levels (or even other games) such as shooting from behind barriers or angling down from a balcony. Similarly, players may need to test variations on these previous methods and devise and test ways to modify their strategies. Or players may need to try to understand how to avoid triggering the onslaught of killer robots when entering a particular room, by understanding the behavior of this species and how the robots communicate, react, and respond. These are all essential skills—transferring knowledge, testing hypotheses, and understanding complex systems—that one can learn through games.

NESTA Futurelab in the United Kingdom has done some extensive work summarizing research in games in education. In the 2004 report *Literature Review in Games and Learning*, Kirriemuir and McFarlane point to a suite of valuable skills developed through game play that include:

- Strategic thinking
- Planning

- Communication
- Application of numbers
- Negotiating skills
- Group decision making
- Data handling

This may lead one to the conclusion that *all* games are educational. And while this statement may in many ways be true, as evidenced by the work mentioned above, it doesn't get you in the door of a classroom. You can present a game (or call it anything you want) to a group of educators, as I have many times, and tell them that by using this game students will improve their skills in analysis, systems understanding, problem solving, and scientific methodology, and that these skills are lasting and important. An increasing number of educators would readily agree with the value of this approach. They would also all tell you that they'd love to use it *if* they had the time, which they most definitely do not. With the increased demands of standards and standardized testing, as well as the growing syllabus and shrinking class-period length, they just don't have the time to teach all of these valuable skills. What they need is *content*, materials that are directly subject-matter-related, often fact-based, and can be delivered easily to students.

**Content and Process in Games**

Many educators lament the "tyranny of the content," while others secretly (or overtly) like this clear focus. But regardless of their personal preferences, they know that they can't invest any kind of serious time in a pursuit that does not directly address class content. Teachers are somewhat powerless in this situation, as they need to play by the rules that have been handed to them by the state and by NCLB. They don't, however, necessarily need new approaches to address the content *better* than their traditional methods, but rather can't reduce time in which they would normally be helping the students learn content. That is, it may be okay if the new methodology doesn't improve content learning, but it can't make it any *worse*. If the process skills come along for the ride, so much the better, but that alone is not sufficient.

That has led us to the pursuit of creating "content as cover"; that is, activities that bring twenty-first-century (FITness) skills into the classroom under the cover of the content that must also be taught. The content is what gets us in the door, but it is really just a vehicle to teach all of the process skills in a way that meshes with the culture and goals of the learning organization. This approach is extended to what we stress in the design of the games themselves, and the way we assess what learners are learning. For the most part, we just don't want them to learn the content any worse than traditional methods, while scoring better on assessments that attempt to quantify performance of twenty-first-century skills. For learning facts, it is hard to beat didactic teaching coupled with drill and practice for efficiency. For some complex subject matter, we can actually expect them to do better using the old methods. But where games shine is in teaching process skills that are intrinsic to the game while the students are learning the content. Through game playing, students learn how to collaborate, solve problems, collect and analyze data, test hypotheses, and engage in debate. These skills are built into the course of the game activities, and are assessed through research. This is the real "education" we are after through games.

We solved the problem mentioned previously, involving the two teachers, one of whom used games and the other book reports, with the latter complaining to the principal, by having the two classes take the same test. The teacher we were working with wanted to show that his students were doing at least as well on traditional measures as the students doing book reports. This was easily accomplished through the multiple-choice tests that came with the textbook. The students in the game-based class did at least as well on that test, and did much better on the extra questions that were added to challenge the students to think outside the box.

Such learning exchanges—replacing traditional content with games or other innovations—are not always straightforward, which creates a problem in accounting when introducing these innovations. A single day of game play may not do as well as a single day of didactic instruction. But two weeks of game play may do as well as (or better than) two weeks of didactic instruction. An investment often must be made by students in learning how to learn a particular subject matter. Instead of emphasizing learning and remembering particular pieces of information in a given subject matter area, one creates a game that stresses particular expert skills to facilitate learning in that subject area. This has been called "preparation for future learning" (Bransford and Schwartz 1999); in other words, teaching with a goal of preparing the learner to navigate new content as it comes along. There are specific paradigms one can employ to prepare learners for future learning, and it does not mean that one can avoid teaching the content. But it does mean that the content may not be delivered as quickly in the short term. New kinds of learning activities take some time to pay off. Students need to master a set of skills that they can later apply to solve problems and to acquire new knowledge. In games, one can think of this preparation in terms of fundamental transferable skills, such as mastery of game-play controls, mechanics and strategies, and deep domain-specific concepts that, once mastered, can facilitate future learning.

Jumping back to the case with the game-playing and book-reporting teachers, the game-playing teacher fretted for some time that his physics students had fallen behind in their lessons because the book-reporting teacher was well ahead in week two of three for the unit on mechanics. The game-playing teacher's students were preparing themselves for future learning, however, in those first two weeks. In this unit on mechanics the game play emphasized core concepts, particularly vectors, which are a new and confusing concept to many students yet crucial to understanding mechanics. Students also had learned how to take responsibility for their own learning through gaming challenges, and how to experiment with and learn from physical systems. As a result, in week three the game-playing students quickly surpassed the other class, having mastered these fundamental concepts in the first two weeks. Meanwhile the other class had only a superficial understanding of mechanics, and depended on the teacher to provide them with all of their understanding. Without that initial investment made in the game-playing class, the book-reporting students struggled in subsequent lessons.

It is easy enough to look up the word chloroplast on the Internet (you don't even have to leave Google to get the answer to the teacher's Jeopardy-style question mentioned earlier). Merely being able to answer such questions will not make a well-trained scientist or even a well-informed citizen. If you can look up a process on the web or automate it, it won't be a job skill for much longer (if it still is). Students need the other skills we've examined to survive in the twenty-first century. This doesn't mean they don't need content as well. If they have to look up everything on the Internet, then they'll struggle too much to form an argument or interrelate ideas. But if we're going to teach them content, let us teach them content that they'll remember, rather than have to test on Friday so they won't forget it over the weekend. The deep learning that students do in games is more likely to persist than the superficial learning they do through memorization. For now, content is still king, but that doesn't mean we can't continue to use *content as cover*.

# 3   The Aftermath of Math Blaster

Every spring in my general education course we have an interactive session on educational media. Many of my students choose the topic of educational games. Over the last several years the top games have included Oregon Trail, Math Blaster, and Number Munchers.

The chosen games are targeted roughly at the middle grades. My undergraduate students would have been in those grades in the late 1990s or 2000. For those of you who don't know these games, they were originally released in the early to mid-1980s, or about fifteen years before my students played those games (and remembered them as the best of the era).

There are several explanations for why they remembered these particular educational games. I've explored a number of explanations for this phenomenon. My first hypothesis was that schools only have old computers and so older games are what students had the capability of playing. Schools certainly have been upgrade-challenged over the years, and the Apple IIs of the 1980s stayed around in the classroom for much longer than computers do today. Those computers persisted until the early 1990s, as did this software and the five-and-a-quarter-inch floppies it resided on. But the students in my course had Macs and Pentium PCs in their schools. More importantly, my students also used these games at home, where most of them had advanced computers.

I've also considered that these games have kept their names but advanced significantly in design since their original versions. It is true that there have been several generations of these games. We're now on the fifth generation of Oregon Trail, and Math Blaster has surpassed that. Oregon Trail follows nearly the same game play as the version of twenty or more years ago. Math Blaster has changed somewhat: instead of showing the player a math problem to shoot out of the sky (similar to Space Invaders), the game now has a platform style, in which the player has to jump up and grab the answer to the math problem that is drawn on the sky (similar to Mario or Sonic). But the game in most ways is nearly the same as its original form—you get to do something fun and game-like by completing a completely irrelevant math problem.

I also considered that perhaps there have been no other educational games since the creation of these seminal games in the mid-1980s. While the market has not thrived since its heyday (which coincided with its birth), there have been other educational games created in the last twenty years. I do occasionally have students who select games like The Logical Journey of the Zoombinis, The Island of Dr. Brain, or The Incredible Machine. There have not been a lot of other titles, and I'll revisit that point shortly, but there have been other games. Some of these games in fact have sold a million copies, and can be found around the world. Yet they still have not had the success of the others.

The best answer I've come up with is that these games were promoted by teachers and parents alike because it was easy to determine what kids are learning through them. In Math Blaster or Number Munchers it is readily apparent that the child or student should be learning addition, subtraction, and division. There are problems on the screen and kids need to solve them. In fact, Math Blaster's slogan is "Mastering the Basics." It does not pretend to teach anything more challenging or substantial. In Oregon Trail students learned about a slice of history, even if it was just to the extent that the "life of the settlers was hard" (in the words of one of my students in a recent presentation).

Teachers justified the use of this software because it was straightforward to map their use onto curricular objectives. Many of the titles of this era, often deemed the "edutainment era," have been rightfully criticized for providing "chocolate covered broccoli" (Laurel 2001). That is, these games don't provide the player with new kinds of learning, but rather a slightly easier-to-swallow version of drill-and-practice learning. However tasteless this entree sounds to adults, to most kids they'd rather have "chocolate covered broccoli" than plain old broccoli any day (witness the Olsen Twins' "Broccoli and Chocolate" song lyrics, "Now I like brocco-chocoli a lot. Mom, it really hits the spot."). That is why these games succeeded with kids. Knowing that the other choice was drill and practice without the totally irrelevant game play, the affinity for these games becomes even more apparent. It may be argued that Oregon Trail doesn't fit this description because it challenged the students to make a series of decisions. But if you ask kids what they remember from this game, there is one answer that you inevitably receive: "shooting buffalo." The rest of the game play is a notable step ahead of the munching/blasting genre but rarely feels like a challenge to players.

It is notable that these games were the "best" in their genres. Many others tried to emulate their success by trying different candy coatings on other vegetables. Games where you shot things and learned spelling, shot things and read, shot things and learned history, or even just shot things and typed (My colleague Henry Jenkins often cites his favorite, *The Typing of the Dead*) proliferated. Many of these failed because they were, in the words of Jenkins, "about as educational as a bad game, and as much fun as a bad lecture."

In the educational software catalogs of the 1990s all of these games would still easily fall into one of the subject matter classes—Social Studies, Math, Keyboarding, and so on. This is clearly defined content. With the introduction of games like The Incredible Machine, the Dr. Brain series, or the Zoombinis series this direct mapping was harder to make, and these games fell in to the category of "thinking games," which possess many of the same qualities as the twenty-first-century skills set. In some cases, like Zoombinis, subject mappings could be provided to teachers to show how particular math skills were learned through the game, but the connection was harder to see, for math problems were not formatted as they might be on a worksheet or in a book. These newer games were really designed to challenge kids to think in new ways, to apply their current math skills, and to acquire new ones. While some may argue that even these games are not open-ended enough, restricting players to particular goals and/or means, they undoubtedly took a giant leap forward in terms of utilizing the potential of the gaming medium, and established new game designs and linkages with learning that were not previously possible in the pre-computer era. The game play and learning cannot be separated in these games, for the problem solving the players must do is also what they are learning.

The fact that these games were not overtly about content, requiring the teacher to make that connection, made them less likely to be promoted by teachers. During the heyday of these games in the early 1990s, Barbara

Means (1994) wrote: "Although less expensive microcomputer-based exploration programs, such as SimCity or Where in the World Is Carmen Sandiego?, are available, an exploration program is seldom a good patch to a particular classroom's core curriculum and, hence, tend to be regarded as 'enrichment'. As a result, technologies that help students explore various areas tend to have limited impact on students' core educational experiences."

Connecting such games to the curriculum of high school, where classes are rigidly segmented around content areas, is even more problematic. Where in-class time is at a premium, it is difficult for teachers to justify interdisciplinary or transdisciplinary activities that were in "thinking skill" games.

Around this same time, in the late 1990s, the Internet was also being introduced to schools. Now, for the first time, there was competition for the computer room at the school. Teachers had to sign up weeks in advance for their class to have a single day with computers. They couldn't afford to waste that time on "thinking games" when there was so much content on the Internet to be learned.

The final nail in the coffin of learning games was the No Child Left Behind Act of 2001. This sealed the tyrannical reign of content and brought along with it the specter of assessment. Now teachers not only needed to think about addressing a very broad range of skills for students, but also needed to consider the way in which they'd be assessed. The assessments were in the drill-and-practice style of Math Blasters (though perhaps without so much blasting). Efficiency of content delivery and matching to standardized assessment were the respective answers to the tyranny of content and the specter of assessment. When those are the most important metrics, learning games have a difficult fit in the school day. So some of the chocolate covered broccoli games persist, and those that don't overtly put content first have come and gone.

### The Exceptions

At the secondary school level and beyond, the commercial market for educational games has essentially been nonexistent since the mid-1990s. After massive consolidation, takeovers, buybacks, and bankruptcies of game makers, the market for early childhood games still persists. The remaining players include The Learning Company (owned by Riverdeep, which now owns the Houghton Mifflin publishing company), the maker of Reader Rabbit series; Knowledge Adventure, the maker of the JumpStart and Math Blaster series; and Atari Kids (which bought out the Humongous Entertainment line), maker of the Putt Putt and Freddie the Fish lines.

Looking at the product lines from these companies shows two important things. First, the number of new products that these companies have produced in the last five years is in the single digits (mostly fleeting tie-ins to kids' television programs and movies), and most of the titles are at least ten years old. Many of the games are still the original software from five to eight years ago (advertising Mac OS 9 and Windows 95 compatibility), or minor refreshes (some of which still don't run on the most current operating systems). So while these companies are keeping their product lines alive, they are doing little to update or invest in new ones. Second, the really successful products are not marketed as games at all but "learning systems." These products are targeted at parents with disposable income and the desire to have their kids "jumpstart" kindergarten by reading early. While there is little, if any, evidence that reading prekindergarten has any correlation with later success, it is a marketing strategy that has worked.

There is one company that I have left off the list, which has emerged as an educational games powerhouse in the twenty-first century—Leapfrog. Leapfrog started not as a software company, but as an *educational* toy company. Leapfrog's huge breakthrough product was the LeapPad, which combined an interactive physical book with a cartridge, and created "paper-based multimedia." Kids could click on parts of the book and have words read aloud to them, or play simple games using the book. The system has sold phenomenally, and has sustained the company since 1999. Leapfrog later entered the slightly more traditional educational games space with the introduction of the Leapster, an educational handheld gaming system targeted at four- to eight-year-olds in 2003. The Leapster brought classic edutainment to the small screen. Most of the games were essentially simple interactive arcade games from the previous decade updated with educational content. Kids liked them and played them and parents felt good about it since they were learning something. Together these products created a whole "green" (Leapfrog's signature color) section of the toy store, signifying that these toys are good for you too.

Fun and games have managed to survive in products for the preteen set primarily because parents are buying toys for kids of this age anyway, and they feel a bit better about it if the toys they buy are "green" and educational. As kids get older their parents no longer buy toys for them (nor do kids of that age have interest in their parents buying toys for them). Perhaps more importantly, there is a broader and more latent view that as students get older, learning is serious and should not be combined with fun. While I am loath to cite the Wikipedia as a basis for fact, I will cite it as a source of opinion. Under the entry for educational game, there is a description of some games for younger kids and a statement that "past the mid-teens, subjects become so complex (e.g., calculus) that teaching via a game is impractical."

## The Rise of Serious Games

The resurgence of educational games is due, at least in part, to the rise in the "serious games" movement. Serious games may be differentiated from educational games by their focus on the post-secondary market, particularly outside of formal education altogether (also referred to as "training" for particular applied tasks as opposed to "education," which is thought of more theoretically). Serious games have put aside the notion that beyond the teen years subjects are too complicated to learn through games. While serious games has become a term attached to any work in the games-for-training space, it owes its origins to the Woodrow Wilson International Center for Scholars, where David Rejeski and co-director Ben Sawyer started the initiative.

The primary consumer and producer of serious games is the United States military. The military landscape has changed along with the economic one. Military personnel need to be prepared to enter a variety of environments, cultures, and situations. They must to be able to understand their surroundings, communicate with familiar and unfamiliar teammates, use new technologies, and make split-second decisions. This requires a new type of training and games have been a welcome addition to the training arsenal of the military.

Among the most well known of the serious games are two games commissioned by the U.S. Army, which were released to the public. The first, Full Spectrum Warrior released in 2004, is a tactical battle game developed by Pandemic Studios. It was originally developed for military consumption but was later released to the general public on a number of PC and console platforms. Full Spectrum Warrior generated a controversy over whether the project was worth the investment the military made in the product. Allegedly the graphical realism was insufficient to train personnel. In brief, people questioned the ability of the game to actually train soldiers as opposed to merely entertaining them and the general public. What is interesting about this controversy, in the context of this book, is that it centers on the learning outcomes of the game. In the end the metrics for measurement were not adequate to determine whether or not it met its objectives.

The other great public military investment in serious games was America's Army, released in 2002. America's Army is a tactical shooter game targeted at recruiting personnel for the U.S. Army. The objective, as much as the content, has caused quite a bit of controversy over the years since its release (and rerelease in version 2.0). While the army has clearly felt the game met its recruitment goals (as evidenced by the contracting of version 2), it cannot provide concrete evidence of the game meeting its goals. There has been analysis of the number of recruits who have played the game, but without controls it is hard to measure the exact impact of the game. Therefore the success of this project is impossible to measure.

Many other military-oriented games have been commissioned more recently, and the U.S. military is inevitably the largest presence at the Serious Games Summits that have met since this movement's inception in 2002. One could argue that this is merely a match of convenience, that the military has capitalized on this medium simply because the medium is most well known for games with shooting at the center of the game dynamics. In fact, the connection is much more significant than that. The military needs motivational tools that can teach complex interactions in a way that will be easy to transfer to real scenarios. Games fit that description perfectly, and it is for those reasons that games can have a much wider application outside of the military space.

While the military may still be the driving force in the serious games movement, games have taken hold in many other disciplines as well. The Serious Games Initiative itself has spawned two sister initiatives, one in health-related games (for medical and public health training), and another in games for social change. One of the earlier serious games projects was Virtual U, a game about administration in higher education. The serious games movement is growing rapidly, with new conferences, organizations, websites, and companies dedicated to this method of training every day. This rapid growth can be attributed to a combination of improved understanding of the power of this medium, along with a changing workplace that constantly demands new skills. As industries change and new personnel stream across them, businesses must find ways to train a large portion of their workforces. The skills with which someone enters a job have a finite lifetime and will need to be upgraded in time. Sometimes this is training to address new technologies, but it may also be due to changing roles and relationships because of shifts that technology has enabled. Regardless, the training needs are rapidly growing.

For the same reason the military has embraced this medium—to enlist motivational tools that can teach complex interactions in a way that will be easy to transfer to real scenarios—businesses too find value. Currently nearly every industry is at least exploring the use of serious games in training personnel, whether on the job or in preparation for careers.

## Seriously Educational Games

Meanwhile educational games have had a resurgence, though without the significant funding that serious games have brought with them. The focus of this second coming of educational games is on creating games that draw upon the best of what modern video games have to offer, tightly linking this to the educational outcomes they are intended to deliver. The game play and the learning need to be inseparable. Today's educational game cannot be layered like chocolate covered broccoli, but rather must be one thoroughly mixed, tasty, and nutritional treat.

Henry Jenkins started the Microsoft Research-funded Games to Teach initiative, which I later joined and morphed into The Education Arcade. Through this initiative we have designed games spanning just about every grade level and discipline, and we have created games in physics, history, environmental science, government, literature, math, and writing—to name a few. Similar organizations have sprouted up at universities around the country, sparked by the renewed recognition of the field, and fueled with some resources coming from the U.S. government to explore the impact of games on education. Academia has led the way, since higher learning institutions don't necessarily need to figure out the economic sustainability before developing products. There have, however, been commercial ventures reentering the space as well. I have followed the tales of Muzzy Lane software, makers of the history game Making History, which has sold at both the secondary and collegiate levels. Other products are starting to come in other disciplines as companies begin to figure out the market. When companies start to see the changing market, a clientele that is steeped in games, and a set of skills to be developed that closely match with the medium, these ventures will increase dramatically.

Educational games are beginning to grab even more attention as a result of the 2006 report published by the Federation of American Scientists (FAS). The report, which resulted from a meeting of experts held in the fall of 2005, calls for greater resources for research and development in the educational games space. The FAS identifies several key reasons why games should play a more predominant role in education that echo many of the arguments from the serious games space—mastery of applicable skills such as problem solving and decision making, suitability for practice in ways that provide valuable feedback to the instructor and learner, and the fit with today's "digital natives." At the same time, the FAS points out barriers to advancement in educational games that inhibit their development and use, many of which focus on the marketplace, such as the high cost of development and the dominance of conservative textbook companies in the education industry. Other barriers include access to technology in schools and the mismatch of important learned skills and traditional assessments. While the report does not necessarily offer new advice to those in the industry, it has served to catalyze interest in a broader audience.

This research and development in the United States is just one component of a worldwide exploration of games in learning in countries including Japan, Korea, the United Kingdom, Singapore, Canada, Sweden, Denmark, and many others. Perhaps the largest educational games initiative is taking place in the United Kingdom as a collaboration between NESTA Futurelab, Electronic Arts, Microsoft, and Take-Two studying the impact of commercial off-the-shelf games on learning in students across the UK (Sandford et al. 2006).

## The Next Generation

So what does this mean for the next generation of learning games? With the latest advances in genetic engineering can we accomplish something like making broccoli that kids actually love to eat (or similarly chocolate with the nutritional benefits of broccoli)? That is, can we make great learning games that are both engaging and rich with learning opportunities? There are certainly challenges that we must meet in respecting the content and assessment needs of present-day classrooms, while integrating challenging and open-ended game play that provides deep learning. We do face technical and design challenges on how to make these work. But the biggest problem may be the legacy that the games already discussed have left behind. From the perspective of games publishers this market is dead. It died in the 1990s with all of the "as much fun as a bad lecture" blaster clones and hard-to-classify "thinking games." Too few games succeeded in that era, and the publishers are stuck thinking about that time. From the perspective of consumers, many of today's parents and teachers are stuck with the edutainment model of the 1980s. They fail to see the power of the medium because they haven't witnessed it themselves. If all the medium has to offer is Reading Muncher 2006, why should they bother? The biggest challenge educational games face is putting the aftermath of Math Blaster aside and moving on. Here we go....

# 4 Great Moments in Mobile and Handheld Games

One of the most famous (and my most trusted) news sources has a motto for its handheld edition: "The Onion just got smaller and harder to read." Recreating (or porting) big-screen titles on small screens misses the point of handhelds. Designing *good* applications for handhelds and mobiles requires taking advantage of the platform and the context in which these games are used. In recent years most of the games that we've seen on mobile platforms have been ports of games from their big-screen counterparts (either on PCs or consoles). This is the strategy that Sony chose for its high-powered PSP handheld, evidenced by the top-selling titles for this platform. A recent look at the twenty-five top-selling PSP games on Amazon showed that twenty of the twenty-five are ports of big-screen games. The PSP has a single high-resolution screen with a traditional D-pad controller and a high-powered processor. The games look great, but there aren't a lot of platform-specific features that demand innovation (though some of the ad hoc multiplayer features via Wi-Fi are an exception). The PSP's lackluster sales can likely be attributed to this lack of innovation.

In contrast, we can look to the Nintendo DS, a dual-screen (one touch-sensitive, the other traditional) design with D-pad, touch screen and stylus controls. Its design has spawned a host of new titles. On Amazon one recent week only twelve of the twenty-five top titles are ports of big-screen titles (and Electronic Arts' (EA) yearly Madden update recently launched). The differences are even more dramatic if we look across cultures to Japan, the home of both Nintendo and Sony. This week all of the ten top-selling *console* titles were for the Nintendo DS platform. That includes titles for PlayStation (1 and 2) and the Xbox (360). Clearly there is both innovation on the design side and demand for these titles. And the sales of this platform reflect the innovation and design. Nintendo sold thirty-seven million units of Nintendo DS in the two-year period from 2005 to 2006, a rate approaching one unit per second.

## The Dawn of Handheld Gaming

Let's step back a few years to 1977, the beginning of handheld game design, when Mattel released its first popular handheld game, known simply as Football. It was a very simple version in which you could run the ball forward (similar to the range of strategies employed by many middle school football teams). Your only options were to move three spaces laterally and forward nine yards (the game was later known as Football I when Football II came out with the advanced features of being able to run backwards, to pass, and to advance a full ten yards). The screen itself was a fairly primitive display with an array of red LEDs. There was no kicking, or field goals or penalties, or end-zone taunts.

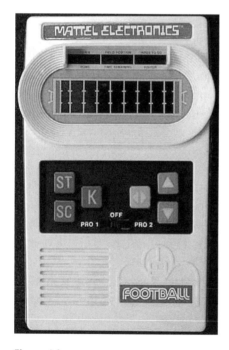

**Figure 4.1**
Photo of Mattel Electronics' Football, courtesy of Rik Morgan of ⟨www .handheldmuseum.com⟩.

If the game play seems limited, it was. In fact you only played offense, and the computer always played defense. Perhaps this was due to limitations in the AI that one could put onto a device at the time. So how does one make football exciting when you can only play offense? Mattel's answer was that the game had to be played with two players, alternating which player was on offense. One person would play offense until either he didn't make first down or scored a touchdown, and then the other player would take over in a similar style.

This meant that you could practice by yourself playing both offenses and trying to run up the score as high as possible, but the real fun and challenge came when you played an opponent. Since you only played on offense, one could only sit and watch as your opponent played. While waiting you could cheer, taunt, distract, or even observe your opponent to try to gain strategies. The minimalist design (albeit from technological limitation more than thoughtful design) left room for a lot of off-computer interaction. I recall sitting on the back of the bus on long school field trips with many of my friends gathered around as we had Football (and later Basketball, Baseball, and Soccer, which all looked and played surprisingly similarly) competitions with each other. When no one was paying attention you might have been able to play for a few minutes, but when friends were around you could play for hours. The one thing that might keep someone playing alone for more than a few minutes would be the hope that later they would be able to show off their skills to friends.

Years later (nearly twenty-five to be exact) Mattel re-released a few of these games in "classic" versions. I was on an airplane and saw the flight attendant take a time-out playing Football. When he saw me looking over his shoulder he immediately challenged me to a game. While I didn't know this person at all, the old Football skills came right back and there was a mix of competition and nostalgia. It wasn't really about the game. I had games on my handheld sitting in the seatback pocket that were almost objectively better than Football in every conceivable way. They had better graphics, more choices to make, greater realism, faster responses, and superior controls. Yet, the tiny red LED lights of Football I were surprisingly compelling.

The fact is that it wasn't really about the game, it was about what the game facilitated. The fact that the game was so spartan meant that the experience centered on the person-to-person interaction instead of the game-to-person interaction. The game simply provided us with some arti-fact to interact around. If we examine this interaction in an activity theory framework (Engeström 1987, 1993; Kuutti 1996), we look at the "activity" as incorporating not just the single player and the computer, even though that is the limit of direct interactions that affect the quantitative outcome of the game. Instead we view the activity as incorporating both players and the evolving negotiated set of goals that they create. In this particular interaction between myself and the flight attendant, we negotiated an out-come where we both tried reasonably hard, but limited taunting and nei-ther of us got beaten too badly. This was quite different from the activities conducted on the bus in my younger years, where the gaming not only included the two players currently playing, but many other friends who were looking on and participating in a variety of ways from "getting dibs on the winner" to encouraging or discouraging the current players.

This illustrates one of the key lessons of handheld design: Handheld games are social—despite the small screen they are meant to be played in public with friends. The design principle (whether intentional or not) of using handhelds to design "activities" that extend beyond the screen has been one of the most successful formulas for handheld games. While hand-held games have evolved significantly, this lesson holds true, and the in-dustry comes back to this theme more heavily after ignoring this for a while. Its relevance to both game design and learning will be seen later.

### The Portable Console Emerges

Mattel continued to dominate the handheld market in the late 1970s and into the 1980s with dozens of variations on the LED sports game theme. The minimalist design leading to interesting activities helped fuel the suc-cess of this platform, as did the lack of competition in the market.

The fall of Mattel's line came at the hands of Nintendo. The rest of the 1980s were dominated by Nintendo's Game & Watch (it's a game and a watch!) line of LCD-based games. The advent of the LCD screen made it possible to create an ecology of games with a greater diversity, at least in terms of the looks of the games. Sports games continued to develop and puzzle games entered the scene. One interesting innovation that the later models of the Game & Watch series offered were dual-screen models, such as the Zelda game series that offered two screens of information delivered to the player. In this context, the second screen was as much an innovation of necessity, given the very limited information the fixed monochromatic LCD screens could provide, as it was creative design. Nintendo would revisit this design years later.

But the big advance here was the porting of arcade games to the small screen. Many of the titles were ported from Nintendo's arcade and console games including the Zelda and Mario series. While the tiny versions hardly replicated the experience of the arcade or full-sized consoles that were introduced later, the characters were the same, as were many of the objectives. This enabled cross-platform marketing, if not cross-platform game play.

While the console/arcade to handheld porting sustained the Game & Watch line for a number of years, more importantly this experience enabled Nintendo to go on and create the Game Boy in 1989, the first handheld console. Game Boy (whose progeny still live on today) was a black-and-white LCD handheld console that shipped with a little game called Tetris. There were many innovations that Game Boy offered, but I'd

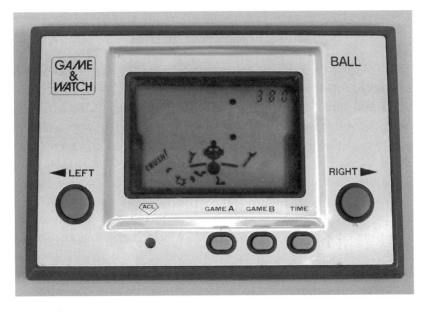

**Figure 4.2**
Photo of Game & Watch's Juggler game, Wikipedia, at ⟨http://en.wikipedia.org/wiki/Image:Game_and_watch_Ball.JPG⟩.

**Figure 4.3**
Photo of the Nintendo Game Boy, at ⟨http://en.wikipedia.org/wiki/Image:Gameboy .jpg⟩.

like to highlight two that are particularly germane to the principles of educational handheld game development.

The first innovation is the inclusion of Tetris. They didn't include a Mario game that would likely only appeal to kids, but rather an addictive puzzle game that was played by young and old alike. In fact, I got my own father a Game Boy, which he used somewhat regularly. This likely gives Tetris the distinction of being the only video game he has ever played. The combination of the casual nature of Tetris and the handheld platform enabled the experience of playing a game at the breakfast table, in bed, or anywhere one might spend a few minutes. As a result, the Game Boy/Tetris combination appealed to adults, even those with little to no interest in video games. The small screen and flexibility of playing place made it feel less socially isolating and awkward. It was at the same time more public and private. The player is more public in being able to play when there are nonplayers around; yet there is no big screen that can easily be observed by others, making the experience more private (and easier to try and fail, or simply to try and not be noticed by onlookers).

The second innovation was the clunky Game Boy Link cable. This allowed two Game Boys to be connected to each other. This feature wasn't used (or at least was not heavily promoted) in many games for nearly a decade. However, when the Pokémon craze hit in the late 1990s, the need for interacting with other people became apparent. Pokémon is a very social game. The card game centers on competing and trading Pokémon characters with others. It is a collecting game. Holding true to the Pokémon style of game play, the link system allowed two players to connect their Game Boys together via a cable to trade Pokémon. The particulars of how this was done changed over the years, but it always involved players having to physically connect their machines, and then interact with each other as they conducted trade. Thus, social interactions explicitly entered into the game dynamics. One might describe the activity here as not only the single instance of trade (social) and game play (solitary), but also as the whole suite of interaction, preparation, collection, collaboration, and competition that players would go through over weeks, months, or years.

The social component of game play on handhelds is a critical innovation in the context of educational applications for a number of reasons. A strong social component for handheld games:

• Allows the game designer to *create a flexible and ever-changing complex game dynamic without needing to specifically program all of the behavior into the games*. This reduces cost and promotes customization.

• Promotes the ability to *adapt games to a number of different styles* such as collaborative or competitive, or a hybrid of the two.

• Provides the opportunity for players to *learn from and teach each other*. It creates a culture around the game where supporting each other's learning about the game itself is valued.

• Creates a situation in which players *learn specialized communication*. Players need to be able to speak with each other, listen to each other, and even write to each other about the games.

• Produces a social dynamic in which players *need to construct arguments, and strategize with and against other players*. These skills are in line with the higher-order thinking skills we are trying to promote.

There is an extensive literature on computer-supported collaborative learning (e.g., Strijbos, Kirschner, and Martens 2004), which also suggests that activities in which computers act as the intermediary between learners can be quite effective in fostering learning for many of these same reasons.

## The Game Boy Successor

For many years the Game Boy reigned supreme. Many other handheld consoles came and went from companies such as Atari and Sega. The Game Boy itself evolved over time, acquiring get a color screen, and appearing in ever smaller models. In total the number of Game Boy variant units sold is in the hundreds of millions. But the game-play experience changed little over the life of that line (which still lives on today).

Nintendo finally dethroned the Game Boy with its own Nintendo DS, which was released in 2004. Despite the release of the more powerful (in specs) Sony PSP months later, the DS has been the dominant handheld console since shortly after its arrival when it overtook the Game Boy in sales. The DS brought back the two screens of the Game & Watch models, enabling one of the screens with touch sensitivity. Also included with the DS were ad hoc wireless networking, a microphone, and a lineup of games designed to take advantage of the platform.

My own experiences with Nintendo DS demonstrate some of the unique aspects of the platform, including technical features and design, as well as how these interface with the portability of the model. Like many "older" gamers (often defined as anyone over twenty-one) I don't get as much time as I'd like to play games. The majority of my opportunities come when I'm traveling, which happens fairly frequently. On one of my trips I had recently purchased the game Into the Blue, a DS game in which you play characters marooned on an island trying to survive. Into the Blue actively includes all of the controls on the DS. You need to dig in the sand using the touch-sensitive screen, move your character with the four-way directional keys (D-pad), and start a fire by rubbing a stick on rocks with the Left and Right keys in the back of the unit. Upon starting my first fire I was told to "blow" to keep the fire going. I looked for the "blow" button in vain and my fires died again and again. Then it occurred to me—was I really supposed to physically blow on the screen? How would it know that I

**Figure 4.4**
Photo of a Nintendo DS, at ⟨http://en.wikipedia.org/wiki/Image:NintendoDS_Warm
.jpg⟩.

was blowing? I looked around the plane and found that most of the people
around me were sleeping, and no flight attendants were nearby with whom
I would raise suspicion. So I blew on the DS. The flames flared on the
screen, but I needed to keep blowing to keep them going. After a significant
effort in the thin air I got the fire going, and I felt triumphant. The triumph
was as much for my success in the game as it was for performing what must
have looked like a ridiculous act in public. In the wrong context this could
be an inhibition. But in the right context, with friends around, these physical actions fill nearby people in on what is going on in the game, drawing
them into the story and the activity.

Some months later on another flight I got to play my new game Brain Age (discussed in greater detail in the next chapter). One of the tasks in Brain Age is to say the name of a color on the screen out loud and the game will recognize what you say. I knew what I had to do here, and fortunately this time the people around me were not sleeping, as I didn't want to wake them since I talk pretty loudly. The game began and I chickened out at first. My fear of negatively affecting my brain age got the better of me in the end, and I went ahead and called out the colors as they were written on the screen. While it felt somewhat strange, I doubt if I was any more annoying than the people who chat on their cell phones just before take off and just after landing. I did get a few looks this time, though most of them were inquisitive.

The market for the DS is quite broad. It appeals to young and old alike. The young segment of the market is highly competitive and relatively easy to attract, but the older segment of the DS market has traditionally been outside of the gaming market generally. A recent study by the marketing group Parks Associates identified a number of interesting categories of gamers, which suggests that the game market is missing the majority of gamers, and this is where the handheld market succeeds.

Power gamers, who are traditionally identified as "gamers," represent only 11 percent of households (though they spend 30 percent of the money on games). Dormant gamers represent the largest category, according to Parks Associates' study, at 26 percent of households. Dormant gamers are older and lack much time to play games due to other obligations. However, they enjoy rich, complex games when they can find the time to play them. Other categories such as leisure gamers, who play casual games, and social gamers, who play games with other people, also make up larger percentages than the power gamers, at 14 and 13 percent respectively. Both sets of these players tend to play games when they are quick, interesting, and social. This has been a missed market, and one that the DS has capitalized on.

Similarly, educational games need to appeal to more than just the power gamers. They need to appeal to a broad array of learners of different ages, expertise, background, and game-play experience. The Parks Associates study shows that there is a strong demand for games from many more categories of people, and if we want educational games to succeed we would do well to think about these broader market segments in creating them.

## Virtual Responsibility

Another breakout success on the Nintendo DS was Nintendogs. Nintendogs extends the virtual pets phenomenon, most directly attributable to the Tamagotchi lineage. The Tamagotchi in and of itself is worthy of mentioning in the context of innovation in mobile games. The Tamagotchi, a small egg-shaped game with a small, simple LCD screen first appeared in 1997. Shortly after you turn on your Tamagotchi for the first time the egg hatches and you have a virtual pet that you need to tend to. Your pet requires food, exercise, changing, and of course attention. As the Tamagotchi owner you provide these requisites to your pet via a few simple buttons on the front of the screen. Much like real pets, the Tamagotchi require care throughout the day. Consequently game play was designed to happen ubiquitously in time and space. Players bring their Tamagotchis with them everywhere, and provide their pets with that click they desired when they needed to be fed or changed.

Tamagotchi's ubiquitous play was a notable innovation. It was of course noted by many teachers who confiscated the games as they beeped and distracted ubiquitously as well (later versions allowed the games to be paused,

**Figure 4.5**
Photo of a Tamagotchi, at ⟨http://en.wikipedia.org/wiki/Image:Tamagotchi_0124 _ubt.jpeg⟩.

avoiding this problem). More importantly Tamagotchi demonstrated the design of a game that could be played casually (you only played for a few minutes at a time at most) while also offering an extended and deep game-play experience, as people developed attachments to their pets, and correspondingly complex strategies for keeping them alive and well.

As much as the Tamagotchi is notable for its style of game play, it is most notable for crossing the gender divide and creating a handheld game that was widely appealing to girls. Perhaps partially due to the style of game play, the physical aesthetic of the device, and the purposeful design around nurturing, the Tamagotchi was quite successful in the female market.

The Tamagotchi itself has evolved through several generations, though the design has held true to the original a decade ago. New features add slightly greater complexity, such as mini-games that offer some solitary enjoyment, and infrared ports that allow Tamagotchis to interact with other Tamagotchis in competition and just for friendly visits.

Nintendogs capitalized on the Tamagotchi design, and its appeal to females to break the gender barrier on handheld consoles. The basics of caring for your Nintendog (available in a variety of breeds) are the same as for many other virtual pets—feeding, walking, cleaning up after it, and so on. But you can also pet your Nintendog using a touch-sensitive Nintendo DS screen to make your pet happy, and even shout commands at your Nintendog using the mic in the DS. These design features make the interaction with your Nintendog more natural and remove the controller barrier that keeps many nongamers from using games. Nintendogs also extended the social component of owning a virtual pet by creating a game mode called "barking mode." In barking mode even when you were just carting your DS around, your Nintendog would "bark" by sending out signals from the built-in Wi-Fi. If another Nintendog was in range it could respond and record the interaction. When you later turned your DS back on you could find out that your dog had a play date while you were out walking about.

Barking mode demonstrated the potential of adding value from the mobility of handheld consoles. While the traditional design of multiplayer games has focused on intensive long-term interaction, this design did exactly the opposite. Interactions are short in duration, and for the most part unintentional. Yet, they are still important.

## The Mobile Console Almost Everyone Has

I was visiting a game studio recently where I got a glimpse at a new take on virtual pets on a mobile platform. The game, Mo-Pets, is produced by Floodgate Studios. Like many other implementations of virtual pets, Mo-Pets requires the player to tend to their pet, including the popular petting introduced by Nintendogs. At the same time Mo-Pets introduced some novel elements of game play through competition. In Mo-Pets you not only keep your pet healthy, but also train it for competitive show by grooming the pet and teaching it tricks. The player can then enter the pet in competitions by region as determined by the player's zip code. Based upon the training of other Mo-Pets in your area you can progress up the competitive ladder. While there is no real-time competition, players indirectly interact through these regional competitions.

What enables Mo-Pets to provide these regional competitions is the fact that Mo-Pets is always connected to the Internet and can rely on connection with a server to facilitate the competitions. This is because Mo-Pets is installed on what many have described as the most widely adopted mobile console in the world—the cell phone. According to many reports, cell phone games are poised to become the biggest platform in the coming years. A much publicized industry report by Juniper Research (Gibson 2006) claims that the industry will grow to over $10.9 billion by 2009, and nearly double again within a few years after that. While nearly 40 percent of that market is projected to be in Asia, there is still a respectable roughly 30 percent in Europe and more than 20 percent in North America.

These projections are based on several market factors, including the sales of nearly one billion cell phones annually (Gartner, Inc. 2005) worldwide, and the rapid growth in the cell phone games market with nontraditional gamers—including older gamers (here defined as over twenty-five) and women. Cell phones seem to be the perfect casual gaming platform. People carry them around ubiquitously, processor power and screen resolution are increasing, they are connected to networks constantly and they are always on. This enables players to game for just a few minutes while waiting in line, traveling by mass transit (hopefully not while driving in a car, though I have seen this), or walking down the street. The design of most cell phone games fits this market. The top-selling games are often variants of Tetris, Bejewelled, or solitaire. But with the growth in market comes innovation.

The cell phone game market is different than other video game markets in several fundamental ways. First there are hundreds of different cell phones for which designers must write games. Unlike the console market, which is highly standardized, or even the PC market, which is somewhat standardized, the hardware and software support across cell phone models is highly variable. On my visit to Floodgate I was immediately greeted by the wall of cell phones, containing hundreds of models lined up by carrier. In conversations with cell phone studios around the world I have been quoted somewhere in the range of three hundred to five hundred different versions of each game that the designers must create for different cell phones. While technology is improving and newer phones share more common features, they are still quite variable in screen size, supported features, built-in languages, and access to hardware. Some of the porting across platforms is simplified by porting tools and portable technologies like Flash Lite and Java (J2ME), but a substantial effort is still required to get a game onto a significant number of devices.

Second, the distribution model of cell phone games is controlled almost entirely by a small number of service providers. Unlike games for PCs which can be downloaded from a number of places, or purchased on physical media from stores and online, or console games that now have similar purchasing means, cell phone games are almost exclusively purchased from each carrier's "deck" via onscreen menus that take you directly to the carrier. The carrier decides which games are sold through its site, and most importantly where those games are positioned on the site. Because cell phones are not easy to navigate, most casual users will only select games that they find on one of the first few screens. Developers try to get their product on one of those pages, but it is highly competitive and expensive. Without one of those top slots, sales are difficult. While people can search on or navigate to a specific game, the player must really want a particular product to get there.

Unfortunately this carrier-controlled model isn't conducive to innovation. Carriers want to put games with immediate mass appeal on those top slots for fear of not selling any games at all. Looking at those decks shows perennial classics including card games, Tetris, Bejeweled, and a few ported franchise games like Madden from EA. Despite this inhibition of innovation, and the lack of standardization, games for cell phones are slowly innovating and breaking into new spaces. Floodgate, for example, released one of the few multiplayer games in the United States. Pirates of the Caribbean Multiplayer (connected with the Disney movie's release on DVD) puts the player in charge of a ship sailing the Caribbean in search of treasure. While sailing, the player can encounter the ships of other players who are currently sailing and must engage in battle with them. The game has met with quite a lot of success in its release, despite two of the problems that plague the multiplayer cell phone game industry—network latency and small communities.

Multiplayer games that take place on PCs and increasingly on consoles are connected to fast networks with latency times (times to get a message from one player to the other) measured in thousandths or hundredths of a second. On cell phone networks this latency can reach whole seconds. This can be a problem if you are playing a game that demands rapid responses, but it can be worked around (or even become a game-play element) through clever design. The second problem of small communities is indeed an issue. When the market for multiplayer cell phone games is limited, the probability of having a critical mass of players online at any one time may be quite small. Again developers need to design around this, enabling compelling play both with and without opponents.

Glancing through the International Mobile Gaming Awards list of nominees for the most innovative mobile games of the year in 2006 shows a small number of multiplayer (both synchronous and asynchronous) games hitting the market around the world. IMO: The World of Magic is a popular massively multiplayer online (MMO) game in Korea, in classic MMO quest style. The developer, Com2uS, claims that they have developed a solid platform that will facilitate the development of additional MMOs in the future. Blades of Magic by Fishlabs takes the PC experience one step further with the introduction of a 3D MMO on a cell phone. Yet another multiplayer game, Street Duel, included what I think will be an increasingly prevalent component of cell phone-based multiplayer games—a PC counterpart. As one way of addressing the issue of critical mass, cell phone games can provide a PC counterpart allowing players to participate on the go or at their desks, opening up a much larger community.

This notion of cross-platform gaming has not gone unnoticed by the industry at large. Microsoft's Live Anywhere platform is based on the premise of gaming crossing over from the PC or XBox to the cell phone. But Live Anywhere is not about shrinking down XBox titles to cell phone screen size; instead it is about some participation that suits cell phones on that platform, and a different mode of participation on an XBox. Live Anywhere provides the player with a persistent gaming identity that spans PC, XBox, and mobile platforms taking with it lists of buddies, high scores, and points. So while the game may not be the same across platforms, the gamer maintains an identity and artifacts that span multiple platforms.

Additional innovations in cell phone games are starting to take advantage of hardware features. Dance Star is a cell phone take on dancing games. In most dancing games the player needs to keep pace with music using keys on the screen. This is a far cry from the full dance pads of the popular Dance Dance Revolution (DDR) which provides a full gridded dancing platform that allows players to dance naturally as part of the game. Building on the DDR motif, Dance Star uses the camera built in to many cell phones as a means of photographing the player's feet. The dancing is done by trying to overlap your feet with the feet on the screen. While the use of the camera can be problematic due to nonstandard software access across manufacturers and carriers, it certainly represents one way of overcoming the limited interface of cell phones.

**Locative Games**

Going back to the list of innovative games shows another game with an exceedingly simple interface. The game Triangler challenges players on two teams to capture each other by enclosing them in triangles consisting of three players on a single team. There are no keys to use to control your character in Triangler. Instead you line up with your teammates by physically moving around. Triangler, developed in Holland, takes advantage of the GPS information on most cell phones to have players use their actual locations as a means of game play.

In many ways location seems like the most natural interface for mobile games. However, its use in mobile games has been quite limited due to the way that GPS information is provided (or in many cases not provided) to the software layer in cell phone games. First, the GPS information on most cell phones is of the type known as Assisted GPS (AGPS). Unlike the GPS devices in cars or mobile GPS units that receive and interpret signals from satellites to independently calculate position, AGPS must use additional information and processing from a server to determine position. For cell phones that means that the position information does not directly appear on the phone but rather must also use services from the cell phone carrier. Consequently the carrier controls the flow of that information, which comes at a cost. Secondly, the interface for accessing that information is variable among cell phones. While it can nearly always provide location information to the carrier for emergency purposes, it cannot always make it available to software on the phone. Looking at the location-based services for carriers or the requirements for location-based games will show a very restrictive list of devices on which the application will run.

Still, there have been a small number of commercial location-based games (alternatively called "pervasive games," "hybrid reality," or "augmented reality"; see chapter 7). The early pioneer in the space was It's Alive's BotFighters game that was played across Northern Europe. In Bot-Fighters, players entered a medieval playing ground in which they had to virtually track and shoot each other in battle. They could fire at a distance knowing the heading and location of their opponent. Tens of thousands of players in the game's home country of Sweden, as well as in Russia, China, and other parts of Europe eventually joined the game. BotFighters had a surprisingly long run, beginning around 2001 until about 2005. It's Alive created other location-based games in the interim period, but as a sign of the struggle for commercial success in this space. It's Alive has been incorporated into another mobile gaming company specializing in—online poker.

Other companies have tried to enter the location-based cell phone game market with varying degrees of success. Given the market dynamics and software challenges, most of the games have not been very widespread. The Shroud, many years in the making, is now challenging this space again. The Shroud is part adventure quest and part sim farming in the style of Animal Crossing or Harvest Moon. In what should prove to be a highly immersive experience, The Shroud uses location to establish neighborhoods and home territories in which the player is responsible for farming and maintenance, but may travel out to visit other neighborhoods and areas on quests. This interesting mixing of genres, which have appealed to a broader gaming audience with titles such as the aforementioned Animal Crossing and Harvest Moon, as well as quest games such as the Zelda series, may provide the style of game play necessary to broaden the appeal of location-based gaming.

## Location-Based Research

While developing the commercial space for location-based games has been a challenge, a number of research laboratories around the world have been exploring the dynamic space of location-based games. The Mixed Reality Laboratory, headed by Steven Benford at the University of Nottingham in the United Kingdom, has been a pioneer in the field. In collaboration with the performance art group Blast Theory, Benford's team has created a series of location-based games using PDAs, cell phones, PCs, and live actors. "Can You See Me Now?" was the first of these games, followed by Uncle Roy All Around You. In Uncle Roy players need to find and meet Uncle Roy within sixty minutes. The players must unravel a series of clues and follow directions provided to them through the devices, but also through actors that are currently engaged in the game. Not all of the information is trustworthy, however, and players must decide which instructions to follow.

Other games in this genre include Play Research's Pirates in which players use location-aware PDAs to role play as pirates, and Playbe's Mad Coutndown in which teams of players track down and defuse a bomb in a building, also using location-aware PDAs. The advantage of PDAs or cell phones in this genre of gaming is the ubiquity of the devices. However they have a trade-off in terms of screen size and the ability to convey virtual information in real space. Another take on location-aware gaming has been led by Adrian Cheok in Singapore's Mixed Reality Lab. Instead of PDAs they have used head-mounted displays and computers to present a more immersive virtual component to the experience. Human Pacman is the most famous game coming out of this lab to date. As one might imagine, Human Pacman challenges players to become Pacman, running around a real space gobbling pellets that are visible in the head-mounted display, and also trying to avoid ghosts, also only visible through the head-mounted display.

Location-based mechanics have tremendous untapped potential in the serious and educational games space. There is no better way to convey to players a sense of authenticity in learning games than to have them actually participating in real space. There are no doubt challenges in creating such games, pertaining to their scalability and transportability, but those barriers can be overcome.

## Alternate Reality

Games that blur the line between the real and the virtual need not be laden with technology. An increasingly popular form of gaming, known as alternate reality games (ARGs), use minimal technologies to create an experience that embeds games directly in player's lives. ARGs exist right on the edge between fiction and reality, real space and cyberspace. You can't define the playing "space" of an ARG, for it typically exists as a connection of integrated resources and media that bridge across the real and virtual world. What does define an ARG is a mystery or puzzle to be solved and the means, however complicated, to solve that mystery or puzzle, eventually.

The first widely know ARG is known as The Beast and was launched in 2001 in conjunction with the move *AI*. It was a subtle launch. In the credits of the movie was one obscure reference to a Sentient Machine Therapist by the name of Jeanine Salla. Those who Googled this name took the first step in uncovering the mystery of Evan Chan's murder. Through a series of Web sites, phone calls, and real-life events that unfolded over months, players in this game began to unravel more of this mystery.

Perhaps more interesting than the game itself (and certainly more interesting than the movie) is the emergent community that developed around this game. Calling themselves Cloudmakers, this group, which totaled around 7,500 active users (MacGonigal 2003) at its peak, assembled to help solve the mystery of The Beast. They were quite successful and integral to the ultimate solution of the mystery.

While there have been quite a few other ARGs since The Beast—including the failed game Majestic, launched by EA as the first pay-for-play ARG, which met with an untimely demise around 9/11 when the distinction between game and reality blurred a little too much. The most famous ARG is likely I Love Bees, launched by Microsoft to promote their new game Halo 2. This game, again unfolding through many media, was

math problems, and (at best) trying to say the color of the text of words, rather than reading the word itself (i.e., the word "red" written in blue should be said "blue" not "red").

Yet at the 2006 edition of the Edge Awards, which recognize innovations in video games, it was indeed a game in which counting as fast as you can in order to measure the age of your brain, called Brain Age, by Nintendo, which won out as the best video game of the year. Now granted this isn't the Academy Awards, or even a video game award that recognizes the hugely successful commercial titles from the big publishers. But the Edge is an award that is granted to titles that push the envelope of video games, and may indicate what is coming down the road.

The success of Nintendo's Brain Age both at the 2006 Edge Awards and in sales is notable first for the platform on which it is played. Brain Age is not played on a TV-based console or a PC, but rather Nintendo's DS portable. The DS, as mentioned in the last chapter, is on its way to becoming Nintendo's best-selling platform ever (though the novel Wii is also doing quite well). It doesn't have flashy graphics, but it is portable and provides for input via touch screen, stylus, voice, and the usual D-pad. Brain Age fits well on this platform (though you do have to turn it sideways), allowing for spoken and written tasks. This demonstrates the novelty of design in mobile game playing. The fact that a mobile game won the Edge Award (and that there is in fact another separate award for mobile games, reserved for cell phone-based games) demonstrates that mobile games not only are on the rise, but also may indeed become the dominant form of video games in the future.

The second breakthrough is that Brain Age is a distinct departure from most recent video games. It resembles neither the mass media hits like Grand Theft Auto or World of Warcraft, or the EA franchises like Madden or FIFA, or even Nintendo's popular lines like Mario or Zelda. It is stark. It is simple. And some would argue it is educational. Equally important, and perhaps because of these very features, it also appeals to users outside the typical eighteen- to twenty-five-year-old male video game-playing audience. Brain Age appeals to baby boomers and the thirty-plus parents who grew up on video games (and no longer have the time to play them). It advertises that it can be played in "minutes a day," something that appeals to these older and more casual gamers. It also uses input that is more natural—a stylus and voice, rather than the usual twitch controls like the four-way D-pad. It can be played on the couch or at a desk (as long as you don't mind people walking in on you while you're saying "red, black, yellow, red"), rather than in front of a fixed screen in the living room. And most importantly, for these busy folks, it is "educational," justifying some time spent playing instead of working.

Brain Age doesn't try to teach you facts about science or history, but instead claims to reduce the age of your brain (to that ideal game-playing age of twenty, as defined by the game), through exercises that activate unused portions of your brain. The pitch is very convincing, with pictures of "low activity" and "high activity" brains. The marketing seems very scientific, and is supported by the work of Dr. Ryuta Kawashima, a prominent Japanese neuroscientist. And people are buying it, despite some studies that do not show any marked improvements in cognitive functionality with respect to activities outside of the game.

Brain Age is the staple of Nintendo's Touch! Generations line of games, which also includes Big Brain Academy and even Nintendogs. The tag line on Nintendo's Touch! Generations Web site is "You don't need to know the rules. Just touch and go." Nintendo emphasizes that these games are quick, fun, and easy to learn.

The success of Brain Age represents the convergence of two important ideas—the rise of the mobile platform and the rise of educational gaming. Educational mobile games are not only the intersection of two small markets but also a huge growth area at the interface of two of the hottest ideas in video games. It may be what saves educational games from the aftermath of Math Blaster.

Brain Age got at least part of the recipe (or a recipe) for mobile learning games right. The game play and learning goals, however, are quite modest. The brain-training exercises are clever, but have rather limited application. It is what my colleague Scot Osterweil (codesigner of The Logical Journey of the Zoombinis) calls "mental calisthenics." There isn't a lot of deep, specific learning in Brain Age, but it feels good to stretch your brain just a bit.

Creating the right recipes for mobile learning games presents a greater challenge. While the field is quite new, it can draw upon related fields to define some starter approaches. This includes drawing from the combined knowledge pools of education, gaming, and mobile devices. Tapping into the successful strategies of video games, and in particular those of mobile video games from the last chapter, is one way to bootstrap the creation of mobile learning games. But we must also look to the learning technologies literature to identify the successful strategies of classroom technologies, and beware of the barriers that many of these technologies have faced in achieving widespread adoption.

## Learning from the Learning Sciences

One of the great challenges of education is the problem of transfer (Bransford, Brown, and Cocking 1999). Transfer is the task of taking knowledge or skills learned in one context and applying them to another. For example, a simple transfer might involve students learning addition through word problems in class later applying that skill in the grocery store to figuring out how much fruit they have bought if they got two apples and three oranges. In that case we may expect the knowledge to transfer from the classroom to the store, as the tasks in the word problems and grocery store are likely to be quite similar (referred to as *near* transfer, whereas application in a much different context is known as *far* transfer).

It turns out that people in general are quite bad at transfer (Bransford and Schwartz 1999). In many cases, people are unable to take what they have learned in the classroom and even apply it in a slightly different context within the classroom (e.g., transferring from one type of word problem to a different type of word problem). Yet there is little value to learning that can only be applied in its limited initial context. It is highly desirable to think about ways to apply learning more generally. Designers of educational materials face this challenge and often try to design strategies that promote transfer into their curriculum, activities, or technologies. Some questions that must be answered in seeking to apply games in an educational space are, "What is the evidence of transfer from video games?" and "How can we promote greater transfer from games to other tasks?"

Jim Gee (2003) points out that games are actually quite good at promoting transfer, challenging the players to apply what they have learned in one part of the game (e.g., safely targeting opponents by shooting around a corner) in another part of the game. Often skills learned in one game may even be applied in another (e.g., running between a giant's legs may be a good way to escape). Transfer outside of the game to other non-game-related tasks is harder to identify. But one of my favorite examples of transfer from within a game to another context comes from the world of alternate reality games (ARGs as described in the last chapter). The example concerns Cloudmakers, the massive group that formed on line to face the challenge of The Beast. This example is interesting because it demonstrates the potential (and the limitations) of transfer of quite abstract and complex skills.

Shortly after the events of 9/11, Cloudmakers gathered in their traditional forums to "solve" the mystery behind the tragedy of 9/11—who were these hijackers and why did they do what they did (McGonigal 2003). They decided that the skills that they had demonstrated and built during the playing of The Beast—cybersleuthing, puzzle solving, collaborative teamwork, code cracking—could be used to shed some light on these events as well.

Ultimately the group used some of these skills to decide this new pursuit was a little nuts and that this solution was better left to the experts. This scenario introduces one of the problems of learning through authentic role play—one needs to note the limits of knowledge and expertise. Role playing a doctor to learn more about genetic diseases can be a powerful design for learning, but doesn't qualify the player to actually consult on genetic disorders.

However, this kind of near role play does have a great potential for learning. Games that are situated in the real world indeed have a firm foundation in a number of powerful learning traditions and sets the bar for new learning styles demanded in the twenty-first century. We may do a lot to promote transfer if we can in turn learn from these kinds of games and learning traditions such as the following:

### Collaborative Learning

Games that are situated in the real world with mobile devices may involve competition, but are often designed around the theme of collaboration, involving coordination across multiple groups through time to solve problems. Since they are situated in the real world, players can integrate real skill sets and appropriate tools that help to solve problems and foster communication (e.g., IM, phone calls, message boards, etc.). Collaboration is modeled in an authentic way—the problems they are presented with are large and require multiple people with varied skills.

### Problem-Based Learning

Games are often based around challenges or problems. The problem can vary in how fictional it is—allowing players to bring to bear (or foster the

in this space, describe the implementation of the game and then evaluate satisfaction and usability measures. They stop short of evaluating actual learning outcomes, which require further refining learning goals, establishing metrics, and scaling the studies to larger numbers.

One controlled study of learning games was conducted on the Skills Arena game, a simple math game played on the Game Boy platform (Shin, Norris, and Soloway 2006). Skills Arena is a simple skill-and-practice game in which students solve math problems. In fact the sense in which it is a game is somewhat limited to the platform that it is on, the graphics, and the fact that the player has an avatar (a graphical representation of the player as a character). But otherwise it is strikingly similar to flash cards, the control against which it was compared. In this study they found that playing the Game Boy version (as opposed to the paper flash cards) improved scores of the second graders on standard math tests. They further found that playing more often was correlated with higher scores, and that playing on the Game Boy significantly improved attitudes towards math compared to the flash card group. So while the game play may be minimal, it still was associated with positive learning and attitudinal outcomes, a promising result for learning games.

### Designing Mobile Learning Games

The rest of this book describes our forays into mobile learning games, outlining the principles of design, along with the outcomes of implementing these designs with a variety of audiences. Chapter 6 describes how mobile games can adapt to the constraints of classrooms through the use of participatory simulations. Chapter 7 shows how the real world can be effectively mixed with the virtual world in augmented reality games. Chapter 8 further explores augmented reality games and how they can effectively be tied to the geographical locations and assets. Chapter 9 describes the use of role play and collaboration in mobile games. Chapter 10 details how authentic feedback in mobile games scaffolds learning and game play. Chapter 11 shows how real-world problems and debates can be integrated into mobile games. Chapter 12 is about mobile games that can be played anytime and anywhere, such as our game Palmagotchi. Chapter 13 concludes with some possible future directions for mobile learning games.

# 6 Participatory Simulations: Technology Adapting to the Classroom

Consider the following scenario:

"Okay class, is everyone here? We're still waiting for two more. We'll give them a few more minutes before we leave for the computer room. Okay, they're here, let's go."

The students head toward the computer room at the other end of the school. They arrive but discover the room is locked and unstaffed. Fortunately the teacher across the hall has the key and lets the class in only to find that the room has been rearranged since the class's previous visit, to accommodate more computers, at the cost of access to the white board. Already eight minutes of a forty-two-minute class have gone by.

"Everyone find a computer and log in with your school ID and password."

This command from the teacher is followed by the usual barrage of responses from the class:

"I can't remember my password."

"My ID isn't working."

"This computer isn't on the Internet."

"Mine has a blue screen saying some weird stuff."

Having become accustomed to the situation, the teacher calmly says, "If you're having trouble getting on to a computer, just pair up with the person next to you. You'll have to go to see the technology coordinator who is here on Tuesday and Friday afternoons to recover your password. Please try to get this done before our next class in the computer lab, which isn't for four more weeks, so it shouldn't be hard to find the time."

The students settle down and tax the capabilities of the network as the computers slowly load student profiles. Thirteen minutes have now gone by in the class period.

The teacher continues. "Remember a couple of weeks ago we were talking about Mendelian genetics? The lab we'll be doing on the computers today will use what we learned in those classes. So go to the menu on your computer, find the science section, and launch the application called Mendel's Lab. Then look for the item called Crossing Cats."

Slowly the students fumble through menus. Some complain that the application isn't in their menus, and log off and join one of their neighbors.

Finally the applications launch and so do the complaints:

"I don't see Crossing Cats in my menu," says one student.

The teacher replies, "Let me see. Shoot, this is version 1.5 and we need version 2.0. I gave the new version to the tech guy six weeks ago and it hasn't made it on."

"Fine," the teacher continues, "we'll use Dihybrid Dogs instead. Please disregard the handouts that I gave you and use the on-screen instructions instead."

Finally the class is getting to work, after nineteen minutes into a forty-two-minute class period. That leaves about eighteen minutes of good working time, allocating five minutes to shut down computers and get back to the "normal" classroom. The students work through the activity clicking in silence, except where students have been forced to double-up on computers, in which case they compete for the mouse to control the pace of the activity in a way that they can manage. Most of the students get no more than halfway through the intended assignment and will have to wait weeks for another opportunity to finish their work in the computer room.

People wonder why technology hasn't radically changed education. While there are many roots to that cause, as this scenario illustrates the bane of the computer room is very high on that list. Most of the computer lab experiences that I have witnessed are unfortunately very much like the preceding example. Let's look at what went wrong here.

## The Computer Room

The computer room paradigm is problematic in practice for a number of reasons:

The computer room is in another place.   To use school computers, teachers typically have to transplant their classes from their home classroom to another place. This means relocation time (which is not trivial in cases when periods are often as short as thirty-five minutes). This also means working in a variable, unfamiliar environment. In this case the room had been rearranged, making it harder to teach the way this teacher wanted to.

Access is irregular.   According to the most recent (2005) data from the National Center for Educational Statistics (NCES) (2006), the average student-to-computer ratio in the United States is fewer than four students per computer. That number may seem like a lot, given that a student could use a computer one day a week. But when you account for the "instructional" computers that are distributed one or two at a time in each room around the school, the number that students actually have access to in a group situation drops substantially. Classes are likely able to get time once or twice a month at most in the computer rooms and labs. This introduces the problem of scheduling. Due to the infrequency and irregularity of time spent in the computer room, if a teacher wants to do an activity based on a particular topic in the computer room, available time may not coincide in time with the coverage of that material in class.

Computer labs are not maintained.   Even when teachers get access to labs, the common computers in these spaces are often poorly maintained and updated. This results from inadequate technical staff in most schools. Nearly two-thirds of all schools did not have full-time technical staff in the 2006 NCES survey. The technical staff-to-computer ratio is often a mere 10 percent of what it is in the corporate world (and pay is similarly skewed very low). So computers cannot be maintained at a reliable level. Often the only solution that the limited staff can offer is to lock down malfunctioning computers, preventing installation of new and valuable software, which further reduces the utility of the machines.

Computer labs are not conducive to teaching.   Labs are shared spaces, and typically are designed to cram a large group (thirty or more) of students—along with with computers, desks, boards, and storage—into a room that was originally designed to hold a lot less. Students sit at desks facing their monitors, often along the periphery of the room. It is difficult to walk about, and hard to see what the students are doing. This unfamiliar space doesn't allow teachers to use the skills and modes of teaching that they are accustomed to deploying in their own classrooms.

Computer labs are not conducive to learning.   From the student's perspective the computer lab is also suboptimal for learning. It is hard to talk with anyone but those on either side of you. Getting up and walking around is discouraged, since the room is often crowded and not designed to have more than one person per computer. Students are isolated in the world provided them by their computer screens instead of being a part of a learning community.

It is true that to some extent these factors result from the focus on using cumbersome desktop computers in a computer lab and would be alleviated by using laptops in the usual classroom. There is some great potential for laptops in the classroom, and it is likely that years in the future we'll see desktops and computer labs fade away in favor of mobile labs or one-to-one laptop initiatives. However, there are two reasons why I dismiss this solution for now.

First, implementing laptop programs is still expensive (both in acquisition and maintenance) and impractical for most schools. Laptops, while approaching the price of desktops, have higher operating costs when one accounts for security, batteries, carts, and most of all maintenance. So for now laptops are likely to be out of reach of most school budgets, or at most they only partially solve the problem by addressing the inconvenience of relocating to labs, but not the frequency or maintenance issues discussed above. Second, and more importantly, is that laptops still put students and teachers in a different mode of teaching and learning—computer-based learning. Students are fixed to where their computers are, and there is an expectation that the computer, not the teacher (or the student) becomes the center of the learning activity. It is as much a result of the design of computer activities themselve as it is the form factor, but computer based-learning activities involve spending the majority of time interacting with the computer (or at most others through the computer). Opportunities for real-world interactions with partners and classmates are rare. There is more on this point in the following section.

### Live Long and Prosper

Let's peek at another classroom that is doing a computer-based lab on genetics.

"Okay, class, let's review your problem sets that you did last night on Mendelian Genetics," the teacher begins.

Eight minutes go by as students peer-review one another's work on paper and a couple of sample problems are done at the board.

"Good, most of you are doing quite well," the teacher notes. "But there are still some misconceptions out there. Today we're going to be using Live Long and Prosper, another PDA-based game. Can I have a volunteer to hand out the Palms?"

As the Palms are distributed, the teacher continues: "When you get your Palm, turn it on and launch Live Long and Prosper. Enter your name and wait for the class to be ready."

Next: "Your goal in this game is to live as long as possible and reproduce. Your ability to survive and reproduce is influenced by your genome, so figuring out what the genes stand for is critical in survival. When the game starts you'll see that you have a sequence of five genes. Each of the genes stands for a trait. The shading of the genes (solid, striped, and clear) somehow stands for homozygous recessive, homozygous dominant, and heterozygous at that position. Your current age (which will continually increase), generation, and total score are also displayed. Your screen will look something like this."

The teacher projects the image shown in figure 6.1 on the screen.

The teacher's instructions continue: "You can mate with other players by lining up your Palms and having ONE player hit the Mate button. At this point you will either get a confirmation that the mating was successful or

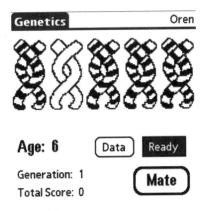

**Figure 6.1**
Screen shot of Live Long and Prosper, showing the player's current genes, age, generation, and score.

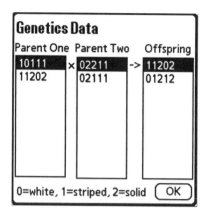

**Genetics Data**

| Parent One | Parent Two | Offspring |
|------------|------------|-----------|
| 10111 | × 02211 | -> 11202 |
| 11202 | 02111 | 01212 |

0=white, 1=striped, 2=solid ( OK )

**Figure 6.2**
Screen shot of data from Live Long and Prosper, showing results of the previous two generations of matings. The codes show the genes of each of the parents and the resulting offspring.

a message saying that you were unable to mate. If you successfully mate, each of the parents will disappear and be randomly replaced by one of their offspring. You can think of these organisms as breeding in discrete generations. After the parents reproduce once, they die. When you reproduce successfully and become one of the offspring, your age will go back to zero, your generation will increase by one, and your score will increase by whatever your age was at the time of reproduction plus a bonus."

Furthermore: "At any point in time you may view the data of your successful matings by hitting the Data button. However, at some point you might die. If this happens you will get a message on the screen, letting you know you have died. You can view your data one last time here, before starting a new game with new genes and zero points."

The teacher concludes: "Your goal is to score as many points as possible. Is everyone ready?"

One of the students inquires, "What do the genes mean?"

And the teacher gives the typical response, "That is a good question. Are there any others?"

None of the other students ask questions, understanding now that they're going to have to figure it out for themselves.

"Okay, then let's begin. Stand up when you're ready."

The students get up, and the teacher gives them instructions to begin. It is eleven minutes into the class, and they have reviewed problems from the night before and begun their computer-based activity.

Students move about the classroom. There is "organized chaos" as they look for partners with whom they can meet. There are frequent comments from students.

"Oh shoot, I died!!"

"Is anyone clear in the fourth position? I'm looking for a *clear* to mate with."

"We can't mate. You must be sterile. Is there anyone with a lot of stripes that I can mate with?"

"I bet if we're too different we can't mate. That is why we can't mate."

After ten or fifteen minutes the teacher calls the class back to order and asks the students to find their seats. Thirty minutes of a forty-five-minute class have elapsed.

"So who had the highest score?" the teacher starts.

"Oh, I forgot about score," replies one student.

"147."

"162."

"184."

No one else responds. The teacher continues: "Okay, so is 184 our highest? So how did you get such a high score?"

"I mated as quickly as possible," explains the high-scoring student. "I also looked for genes that were a lot different than mine. I think that having different genes makes you more successful."

"Good. Who had 162? What did you do?"

"I noticed that the first few times I died kind of young. Like around twelve to fifteen. So I tried to mate before I was twelve. Sometimes that was hard."

"So what did other people observe?" the teacher continues.

"Sometimes we couldn't mate."

"We died at different ages."

The teacher follows up, "What ages did you die at?"

"Thirteen."

"Fifty."

"Sixty."

"Anything else?" the teacher asks again. These comments are all written down on the board under *Observations*. The students copy them. "Why do you think sometimes you couldn't mate?"

"I think it was when we had too many genes that were different from each other."

"No. It was because too many were the same," responds another student.

"I think it was when we were too old. We couldn't mate if we were too old."

"It was the combination of the ages. If they were too high you couldn't mate."

A few more hypotheses are made by the students and written down on the board by the teacher in the *Hypotheses* column.

"Okay, how can we test these hypotheses?" asks the teacher.

"We can analyze the data from our meetings," responds one student.

"What about the times when you can't mate?"

"Right," the student continues. "I guess we'll need to write those down."

"Here is a data sheet. At the top is printed Parent 1, Parent 2, and Result," the teacher says, handing out the forms. "You can use this sheet to write down data from all of your matings. We only have five minutes left, so let's get up, get into some groups and collect some more data. When the bell rings you can leave your Palms on the cart, and we'll pick this up in the second half of class tomorrow."

## Technology IN the Classroom

How was this class fundamentally different from the class conducted in the computer lab? Or even how a class using laptops might be? What is obvious is that it *is* different. Time was used differently. The teacher played a different role in the class. The class experience *looked* different and *felt* different. While there are many factors that could make such a difference between these two hypothetical classes (based on real experiences I've had), there are a number of significant factors in the design and use of the technologies themselves that dictate these usage patterns. Some key factors follow:

First, the class took place in the regular classroom. No one needed to move to another room. The teacher and students were in their normal surroundings. They didn't waste time going somewhere else, or dealing with the logistical issues of being in another room. The white board could be used by the teacher, and the students were able to sit in their usual seats and even use their textbooks if they chose.

Access can be frequent.   A set of Palms for this activity runs less than $100 per unit. Enough for the whole class can be obtained for the cost of a couple desktop machines. Palms are easy to maintain and use. They can be rotated around classes on a small cart or bag so that teachers can use them on their own schedules for part of a class across weeks, or intensively within a few days.

Computer labs are not maintained.   Simple Palms in contrast require almost no maintenance. There are no security upgrades or gatekeepers to prevent software installation. They can be entirely reset and reloaded in a matter of seconds or minutes if need be. They can be rebooted in a few seconds.

Labs are not conducive to teaching/learning.   Forget about labs. Palms can be used wherever the students are. Students can get up and walk around, and interact with other students. They can form groups as necessary and be called back to their seats for whole-class discussion or lecture. The flow of the class is something that the teacher controls, not the technology, and can be modified to whatever the teacher feels comfortable with.

## Technology Adapting to the Classroom

Together, these differences demonstrate one of the fundamental design principles that we've used in designing PDA-based games. The games should fit into the classroom, and allow a more natural flow of the class. I'll call this principle *technology adapting to the classroom*. It stands in contrast to the classroom that needs to adapt to the technology, as was the first case with the fictitious genetics activities Crossing Cats and Dihybrid Dogs. When we do our job right, one should see little difference in the behavior of students or teachers in a classroom using our technology versus a class doing a very engaging nontechnology-based activity. That doesn't mean they won't be learning new and powerful ideas and concepts, but that the ways they'll be doing this are subtly different from longstanding engaging methods of teaching, learning, and playing. These technologies should make it easier to engage students in familiar playful modes of learning that may be less used because they often are thought of as being too rudimentary. But the technology can take these modes of playful learning and make them deep and powerful.

It is often said that there are two ways one can use technology in education—you can do *old things* in *new ways* (i.e., taking notes on a tablet computer instead of a notepad, or writing on a smartboard instead of a blackboard) or you can do *new things* in *new ways* (i.e., create immersive simulations where you can become an electron and see out of its "eyes" in virtual reality or travel back in time to interact with colonists in eighteenth-century America in an online gaming environment). The former category makes up the majority of classroom applications that are in use today—word processing, research, and presentations. And the latter is discussed in some of the later chapters on augmented reality. But some of our technology fulls into a third category, which is doing *new things* in *old ways*. We are having students explore phenomena in ways not previously possible while using a style with which students and teachers are familiar. Through technology-enabled role play and interaction face to face with their peers, students can engage in deep, meaningful explorations of complex topics.

That isn't to say that this is where technology should stop, for there is certainly a place for doing new things in new ways (and probably places for doing old things in new ways, which is what the vast majority of educational technology use consists of—internet research in lieu of library research, word processing papers instead of handwriting). But to ask teachers and schools to change what and how they're teaching simultaneously is unnecessarily attempting to solve two problems at once. We can solve one problem first, the more important one, which is getting them to learn new things, and then later, as teachers get comfortable with the technology, they can transition to new ways of learning.

The principle of *technology adapting to the classroom* is one of the most important for getting into the classroom in the first place. Schools are one of the most resistant organizations to change (Tyack and Tobin 1994; Tyack and Cuban 1997; Cuban 2001). The path of least resistance for incorporating technology, which may be seen as an unstoppable force, has been to relegate it to the periphery and apply these new technologies to do exactly the same thing that schools have been doing for a century, and with intense focus most recently: preparing for tests. Rather than figuring out how these technologies can be used to meet the challenges of the twenty-first century, they're being used to meet the challenges of the twentieth century in slightly more efficient ways. Initiatives that have attempted to ask schools to do new things in new ways have primarily failed (Tyack and Tobin 1994). Changing too many variables at once is a recipe for failure.

The fact that these simulations are done in "old ways" alleviates many of the issues associated with going to the computer room. One teacher who was working with a combination of technologies, some of which used PDAs and some of which used traditional lab computers, stated:

> I would definitely use the Palms in the future for activities in my classroom. It is such a pleasure to have technology that works as well as not having to go through the time consuming administrative procedures to use the lab and the frustrations that go with the above.... It took him (an expert) 1.5 hours to get [the software] up and running on one of the computers, but he said that the other computers that I have been unable to fix even after more than 5 hours of my and another teacher's fiddling, have got various problems including CD ROM problems, so he was unable to do a repair. (From Klopfer and Yoon 2005)

When technology can actually meet teachers where they want to be, it can be successful. Teachers shouldn't have to change the way that they teach and manage their classes just to get started. Changing the way classes are taught and managed demands a whole suite of skills that require additional expertise and development that will come in time, but are not always immediately available to teachers. The professional development that is so often lacking can provide teachers with the skills they need to teach this particular content instead of having them learn entirely new ways of teaching and new content. Then teachers can apply the skills that they have mastered in managing and mediating classes, that is, leading discussions, guiding student inquiry, structuring collaboration, and so on. Today's teachers have those skills, but they need the space in which to use them.

## Designing "Lightweight" Technologies

There is another important design principle that is manifest in our participatory simulations, which I'll call *designing "lightweight" technologies*. Our version of PSims closely resemble nontechnological role-playing activities (think of classic camp games in which you might act out the relationships of plants and animals in an ecosystem). These role-playing games embed people inside of systems, in which the participants act out the behavior of those systems. Participatory simulations add a small touch of technology to those role-playing scenarios, which enables them to become more complex and data-driven (for another take on participatory simulations see Wilensky and Stroup 1999). If one looks at the screen of one of our participatory simulations, one sees very little. The technology itself is very "lightweight." It provides a small amount of information to the user, just enough to enrich a primarily human-based social game. This has two advantages. First it allows the teacher to apply familiar methods of mediating the class as previously discussed. Students spend a small fraction of their time actually looking at the screen of the PDA. In one study that analyzed student behaviors using participatory simulations (Klopfer, Yoon, and Rivas 2004), "looking at the screen" didn't even make the top five behaviors in which students were engaged during the activity. Instead they were writing notes, talking and interacting with other students, analyzing data, and walking around.

The second advantage of this design is that it allows the teacher to structure the game in many different ways. Instead of being locked into a particular focus to the investigation, or to a specific way in which the game is used, the lightweight technology keeps the teacher in charge. It is the teacher who decides how the activity can flow, not the technology.

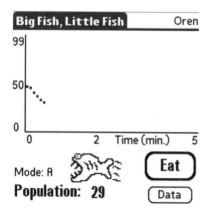

**Figure 6.3**
Screen shot of the game Big Fish, Little Fish, showing the decline in population of Big Fish that have not eaten in some time.

To see how this happens, it is probably useful to outline some of the participatory simulations that we have created. These are shown in table 6.1.

Take, for example, the Sugar and Spice game. This was a game originally designed to teach some simple economics principles of supply and demand. It is loosely based on Epstein and Axtel's *Growing Artificial Societies* (1996). In Sugar and Spice some players are sugar producers and others are spice producers, however everyone needs both sugar and spice to survive. Consequently, the sugar and spice producers trade commodities. The catch is that different people produce and consume sugar and spice at different rates. And in fact the number of sugar and spice producers can be askew. Students investigate what the underlying production and consumption rates are, as well as the ratios of the different types (sugar and spice producers) to determine either the price they should pay for commodities they need or the price at which to sell their commodities. Yet the game specifies few details about what should be taught. Instead it is a lightweight tool that simulates a system of trade. In a workshop on this game, teachers discussed ways in which they had used or would use the sugar and spice game in their own classes (Klopfer and Yoon 2005).

Here the teachers in the workshop discuss the ways in which they could adapt Live Long and Prosper to their own classes and contexts:

Teacher 1:  To me it's a carbon cycle issue. I could say trees are doing... are basically respirating carbon dioxide into oxygen and humans the other way around. So you can say sugar and spice would be a balancing act in that sense. I realize this is going more toward commodity, toward buying and selling, but you can certainly also show homeostasis or stasis of one thing using one and the other thing using the other and you can maintain that for an indefinite period of time. You can show a cycle of... basically a carbon cycle.

Facilitator 1:  Is there way that you could for kids... is there a way that you could go into this and change it to make it carbon and oxygen? Is there a way that you could get into the program?

Teacher 2:  And you could have animals and plants.

The teachers have taken an economics game and adapted it to a biology curriculum. They continue extending the idea in biology and move on to additional subjects:

Teacher 3:  With the way the game is described that pairs would go on with each other that they would just stick with each other and that would be a winning strategy.... I think there are animal pairs that pair up that way 'cause they learn how to stay together.

**Table 6.1**
Participatory Simulations and Their Uses

| Participatory Simulation | Sample Palm Screen | Description | Activities and Adaptations |
|---|---|---|---|
| Big Fish, Little Fish | | Big Fish, Little Fish models a predator-prey system. Some players are schools of big fish that need to eat little fish to survive, while other players are schools of little fish that must avoid the big fish to survive. Players can track their school size over time via numerical readouts and real-time graphs. The challenge is to support as many fish as possible in the pond, and preserve species diversity. | Predator-prey relationships Tragedy of the Commons Fisheries management Animal behavior |
| Discussion | | The Discussion simulation poses a statement for participants to consider, such as: "Technology has succeeded in changing classroom practices." Participants individually rate their agreement or disagreement with the statement and provide a brief rationale. Then everyone must go around and make their case to the other players, and track how their own opinion and rationale change in response to feedback from others. | Science, technology, and society issues Mathematics (estimation) Science (scale) |
| Live Long and Prosper | | Live Long and Prosper is a genetics simulation. Players take the role of an organism with a simple genome (between one and eight genes) that is represented on their screens. Players quickly age and must produce offspring to survive. While players get points for surviving and reproducing, the game quickly focuses on trying to figure out which traits the genes code. | Genetics Ecology Evolution |

**Virus**

Virus is a game in which everyone initially appears to be healthy. Players are then given the task to meet as many people as possible without getting sick. Just how does one do that? That is what players must figure out. As the game proceeds some players get sick. By playing again players can continue to try to determine how the virus works.

Health
Biology
Epidemiology
Exponential growth
History of diseases

**Sugar and Spice**

Sugar and Spice is a simple game of economics, loosely based on the artificial societies discussed in Epstein and Axtell's *Growing Artificial Societies* (1996). In this game sugar producers and spice producers must negotiate trades in order to survive. Along the way players must try to learn how the system works in order to optimize their trading strategies.

Ecology
Economics
Behavior
History
Biology

**NetsWork**

NetsWork is a simulation of social and information networks. Players form connections to others over which they can pass messages. They soon realize which message routes are more efficient, and come to understand the dynamics of the network they have created. Teachers can visualize the networks by downloading game data to a PC using specialized software.

Graph theory
Biological networks
Computational networks
Social networks

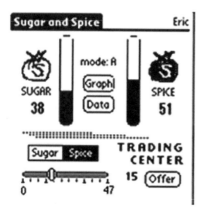

**Figure 6.4**
Screen shot of the game Sugar and Spice, showing the relative quantities of these two precious elements that players must trade for in order to survive.

Facilitator 2:   I've used this in history classes with history teachers before. One of our history teachers was teaching the development of mercantilism in Europe and thought that would be a good time to introduce this. And it gave the kids...it very much personalized it for them.

And some time later in the workshop discussion:

Facilitator 2:   One of the aspects of this [that] I think was most interesting is when I did this in history class. The history teacher and I talked about the notion of this being a how you structure your social system kind of a game. So you could create cartels or you could have price controls imposed by a central government. So you could actually build small societies or subcultures in much the same way that some of you all were starting to have the pairings. You're already beginning to build small structures of society. If you did this over a number of days, you could actually explore really different organizations from that perspective.

Teacher 4:   You know, I was thinking as I was playing this and thinking about media. I kept thinking about looking at large multinationals like Vivendi merging with GE or else looking at how shoes like Nike outsource to smaller companies where it's cheaper to buy and create the materials to send them off that way. So we're going to go over to another country and trade off. So I kept thinking about the social dynamics of it more than the scientific applications.

As this dialogue illustrates, a game that was originally designed to teach economics is readily adapted to teaching on the carbon cycle and symbiosis (the mutually beneficial pairing of animals), or on the development of mercantilism in a history class, to offer only two examples. Similarly the game

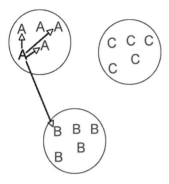

**Figure 6.5**
A sample structure from the game NetsWork, simulating groups of people who know
a closeknit group of other people well (strong links) and other more distant people
less well (weak links).

of NetsWork was designed to teach the basics of social and computational
networks. As in many networks, there are nodes and links between those
nodes. In this case, the students are the nodes and they can make links to
each other. However, they can only make a small number of connections,
and those connections are divided up between those people in their
"group" and those in other groups. The groups are computer-designated
groups of particular sizes that are randomly assigned to students when
they begin the game. Players can make several connections to other people
within their group, and a much more limited number to people in other
groups. The network topology typically looks something like figure 6.5,
which shows players in different groups and the connections that one
player in group A has to players in group A and group B.

Players are then tasked with getting the maximum number of messages
to as many *different* people as possible. They are limited by the number of
messages that they can have out at any one time (i.e., undelivered messages). They are further restricted to choosing only destinations to which
they are not directly connected. So the player in figure 6.5 could choose to
pass messages to only one player in group A (to whom they are not connected), but nearly everyone in groups B (except the one to whom they
are connected) and C.

Most of NetsWork typically is spent creating messages and figuring out
ways that players can get to their destinations. At the end, students are
able to visualize the entire network, including all of the connections in

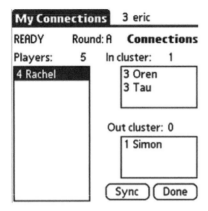

**Figure 6.6**
Screen shot of NetsWork, showing the connections that the player can make at the beginning of the game.

the network, as well as where all of the messages wound up, and statistics on the messages and connections in the system. While understanding networks, an important and emerging science, was the original design goal for this game, teachers have adapted it to a variety of contexts.

Here the teachers discuss how they could use NetsWork in their own classes:

Teacher 1:   Well, in [our math curriculum] we have a section dealing with networks and how to connect them dealing with the best pathway or other pathways.

Facilitator:   So what grade is that?

Teacher 1:   Ninth grade.

Teacher 2:   And when her kids leave ninth grade and come to me in tenth grade, and we work with adjacency matrices and counting the number of paths from one to another and using the power of matrices and stuff like that and so it would fit for her and my courses.

Facilitator:   Other feedback?

Teacher 3:   The analogy that you made about the clusters, I pictured them like two neighborhoods where you've got a lot of people knowing each other in one neighborhood. Why not just in the interface call it, I'm in the neighborhood A. Just change the name to neighborhoods instead of clusters because the game is always going to be played with people and I think the first, I mean especially with kids the first concept of network that they have is neighborhood.

Teacher 4:   I can see this being used in many different sciences, in computer science, in any area where communication is necessary.

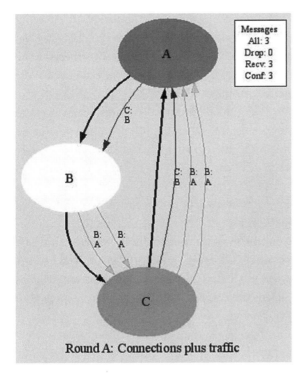

**Round A: Connections plus traffic**

**Figure 6.7**
The graph that is output from NetsWork, showing connections between players and the paths that messages took.

There are many dimensions to the flexibility that this model offers, as demonstrated in this conversation. That flexibility comes in several forms, in content, in grades/ages/levels, and in context and connections.

The flexibility in *content* was demonstrated by the Sugar and Spice game as well. For NetsWork the connections of networks are made to various aspects of the mathematics curriculum and computer science. Other teachers have adapted NetsWork to be used in biology classes (biological networks), social science classes (social networks), and health classes (as models of how diseases were spread).

The flexibility in *grades/ages/levels* is shown in this conversation. Teachers discuss how to use NetsWork in ninth grade math, tenth grade math, and middle grades (the "younger" students referred to in the conversation). The goals and scaffolding that the game provides are primarily provided by the teacher. Therefore, without any technical manipulations the teacher can change the game to be easier or harder, or to focus the discussion on the more or less challenging (as well as more or less relevant) aspects of the game. The game provides the space in which the teacher can create these supports using "traditional" teaching skills.

It is interesting to note that "older" doesn't necessarily mean "better" when it comes to playing these games. Often the playful nature of younger students, perhaps combined with their greater experience in solving game-based problems, make them more successful at these games. For example, I have played the virus game (see table 6.1) with students ranging from fourth grade through university level, and on up to teachers and professionals in the public health profession. The classes that are the "best" (for now, defined as figuring out the most information in the shortest time) are often at the eighth-to-tenth grade level. And the "worst" (taking the longest and finding out the least) class that I have ever encountered was a group of master's-level epidemiologists, containing many experienced adults from the medical profession.

Younger students will play the game again and again. They dive in, playing the game "haphazardly," but develop intuitions (often wrong) based on their experience. As they gain more experience from additional rounds of play they may develop new and better hypotheses and discard some of the old ones through a somewhat informal process of hypothesis testing. As is required in many games (from figuring out how to kill the "boss" in a first-person shooter game, to figuring out how to hold a job in the Sims), the tried-and-true method of trial and error can be very helpful. The trial and error approach (when done systematically) is really hypothesis testing. The students in these age ranges will readily engage in that mode and test and test again. In the end they will need to run some formalized experiments, but those are most effective when based on evidence that they have already collected through play.

As the students get older they develop a notion that the only way to learn is through designed experiments. This echoes the teaching of science in schools (McComas 1996; Sandoval 2005) through which it is conveyed to students that there is a single scientific methodology in which you linearly start with a hypothesis, conduct experiments, and come to a conclusion. After playing the game once the students (high school, university, professional, etc.) will sit down and try to design the single best experiment that can answer all of their questions of interest. They will simultaneously attempt to determine who started the disease, the probability of its transmission, its incubation time, who is immune and why, and how to minimize the number of infections. There are two problems with this approach. First, designing an experiment to simultaneously test such a large number of variables is difficult or impossible. Variables need to be isolated and tested independently through repeated experiments. Second, designing even the simple experiments based on limited experience is unlikely to be productive. Scientists spend much of their time "playing" in the lab, figuring out things about their systems of study through tinkering and small informal experiments. Once they develop some good intuitions they can then start to design experiments. That process is modeled well by the younger, playful students. The "experts" who attempt to theorize in the absence of experience and data will often wind up endlessly spinning their wheels.

There are two other manifestations of the single, linear, scientific-method construct in many of the players of these games—they don't know when to discard hypotheses and they don't understand the randomness inherent in complex systems. With respect to the first issue, it is beneficial to generate and test hypotheses throughout these experiments, but the linear scientific method has provided them with little experience on using feedback to modify or discard hypotheses. Take for example the Virus game. One of the initial hypotheses that students often generate is what I call the "chemistry hypothesis"—that two particular people need to meet to create the initial virus (each containing a part of the necessary components). Once those two specific people meet, they create the virus and pass it around. The students will hypothesize, "Arnold needs to meet with Betty to start the virus," having observed Arnold and Betty were one of the pairings that met right around the time of the virus's overt appearance. They will then conduct a test, find that this is not necessarily the case, and modify their hypothesis: "Arnold needs to meet with Betty after meeting with Cindy." Then they will test that hypothesis, and find out it is also incorrect. Again they modify the hypothesis: "Arnold needs to meet with Betty after meeting with Cindy who has met with David within the last two minutes." And so it goes, on and on, with this series of revisions, the students never knowing when to apply Occam's razor and start over with a simpler hypothesis that could also explain the phenomenon.

As for the stochastic nature of complex systems, while this is something that students have experience with in their daily lives, from weather patterns to traffic jams to the ice crystals on their windows, it is not something of which they have been explicitly made aware. In fact, in many ways the science experiments that they conduct lead them to believe that we live in a deterministic world. These experiments are done once and have a correct answer. If the answer deviates from the correct one this is chalked up to "error." In fact, in complex systems there may be different results in repeated trials due to undetectably small differences in the starting conditions. The Virus game may unfold in quite different ways in repeated trials due to small differences in the initial interactions, or some of the inherent randomness in the system. Additional experience with school science seems to decrease the facility with which students can understand such complex systems.

There is also flexibility in *context* and *connections*. One might call this the flexibility to be creative. The games specify little about their meaning or context. While we suggest certain connections to make it easy for teachers to get started, these are not built into the game, allowing teachers to develop their own scenarios that may be more fun, relevant, or meaningful for their audience. For one teacher this was making linkages to neighborhoods. For another it was the Pony Express or other stories.

Together these flexibilities create a malleable medium—the raw, partially formed materials that can be shaped and bended by teachers. The potential to walk the line between overly structured materials like those that often come with science kits, or packaged software that doesn't allow teachers to exercise their creativity, is the reason many teachers chose their profession. When software does allow that customization, it requires technical modifications, going in to preferences or editing files, which most teachers feel uncomfortable doing. This means that the materials must take a one-size-fits-all approach. On the other hand, giving teachers a tabula rasa or blank slate is unworkable. Most teachers don't have the time or expertise to design entire lessons from pure, raw materials. While this method can be powerful when teachers are adequately supported during the process, this is not a luxury available to most teachers.

So the solution that we have provided here is to give teachers a game framework that can be easily customized through nontechnical means.

This results in teachers' having materials that they have ownership over and matches their needs.

## Personal Engagement

The final design principle that I'll distill from the participatory simulations is *personal engagement*. While engagement may seem an obvious design principle for games, it may not be obvious how one creates something that is simultaneously engaging and educational. Many commercial off-the-shelf games are designed around extrinsic rewards—points or award structures that are easy to measure. In first-person shooters this may be merely points derived from the number of kills, accuracy, bonuses, and so on. In role-playing games this could be wealth, as well as the level of your character. But increasingly games are focused on creating deep, personal attachment to characters in the game, and the rewards are "success" for that character that is often defined by the player themselves.

In these games, as may be the case in others where there are more overt motivators, the reward is "Flow" (Csikszentmihalyi 1990), which means "being completely involved in an activity for its own sake." It is marked by extreme concentration, pleasure, focus, reward, and even exhaustion. Activities that lead to Flow encourage such qualities as clear goals, high concentration, feedback, appropriate challenge, personal control, and intrinsic reward. These same qualities are found in good games.

While many of the participatory simulations contain extrinsic rewards (points or predefined goals for the individual player), these are often just a way to concisely define an initial problem for the players so they can get their minds around the problem. The real reward comes from figuring out the system, which is a collective/collaborative goal. For instance, in the genetics game Live Long and Prosper, there is a point structure that players are given at the onset of the game. The score that you get is whatever your age was at the time of mating plus a bonus according to the following scale: age 21–40 = 5 points; age 41–60 = 10 points; and age 60+ = 20 points.

This scale was designed to get players to explore the upper end of the age ranges, instead of just reproducing as fast as possible (and yes, the notion of "reproducing" gets giggles from middle schoolers on up through adults). In the initial versions of the game there was indeed no score at all. But when we explained to the players that they were just supposed to reproduce and try to figure out the system, they didn't know where to begin. So instead we introduced the idea of points and told the players in the initial round that they were to try to get as many points as possible. This is a tangible goal that gives new players a place to start. The notion of score, however, disappears by the end of the first round if not before. There is the occasional player (usually boys) who will brag about their score, but for most people the goal of scoring has been replaced by the goal of trying to figure out the system.

The reason why figuring out the system becomes a worthwhile goal is multifaceted. First, it is a challenging yet comprehensible problem. Students can understand what they are trying to figure out, even if it takes them a while to figure out how. Second, it is a challenge that they defined. True, it is somewhat inevitable that this is where the game goes, but the way they approach that problem and the specifics of it are owned by them. Third, it becomes a collaborative goal that relies on most if not everyone in the class. Finding the rare clear gene may make anyone in the class an important player at any time.

Finally, and perhaps most importantly, the games are always played from a truly first-person perspective. This is not about "the machine" or even about my "character" or "avatar." The game is about "me." When players are engaged in the Virus game they will often call out "I'm sick" or "who got me sick." In the Live Long and Prosper game, players will be asking "can I look at *your* genes" and "who wants to mate with *me*?" The idea that the games is about "me" is critical to its success. Students are embedded within this simulation game not just physically but also intellectually and even emotionally. The game is not about the machine, but about them, and the machine is merely an extension of them.

The theory of situated learning (Lave and Wenger 1991) states that the learning is integrally connected to the environment in which the learner is situated. As mentioned earlier, an important component of that environment is the other people who make up the community. But it is not merely the learner's mind that is present in this environment, it is his body as well. Others have argued (Winn 2003; Rambusch and Ziemke 2005; Klemmer, Hartmann, and Takayama 2006) that the use of the body in these environments is also critical. The body and mind are integrally connected, as the body facilitates communication, spatial understanding, and reasoning. Thus the use of the body in learning activities is critical.

Designers of virtual environments have learned the importance of creating a virtual experience that mirrors physical reality as close a possible to permit the use of these more natural forms of communication and navigation. The goal of this sense of reality is to increase "presence" (Lombard and Ditton 1997; Lee 2004), the feeling that the person in the virtual environment really *is* in the virtual environment, not in the real one. Some research has shown that increasing such presence is associated with increased learning in virtual learning environments (Salzman et al. 1999; Winn et al. 2001; Winn et al. 2002). Gee (2003) goes further in discussing a principle that he calls "embodied empathy for a complex system," the notion that you develop understanding of complex systems in video games by actually becoming a virtual part of that system. But creating that sense of presence is difficult and we may never be able to replicate the sense of presence we have in the real world.

Typically, in video games that presence comes at the cost of having to design highly realistic environments and characters. In the participatory simulations it comes for free as a result of relatively simple design, so that the players can construct their own meaning about the game and what their role is. Klemmer, Hartmann, and Takayama (2006) state that "interfaces that *are the real world* can obviate many of the difficulties *of* attempting to *model all of the salient characteristics of a work process as practiced.*" The electronic interactions are purposely brief and the explanation is intentionally designed to be about the player. Most of the interactions and interface are nontechnical, as mentioned above, and therefore perfectly mirror real-life interactions.

Attaining that sense of embodiment connects students emotionally to the game, which results in their significant investment in solving the problem at hand. They care about the answer to the problem because it concerns them. I have a particular recollection of using the Virus game in a low-performing school (specifically, it was an urban school for slightly "older" students who had spent some time away from school and in jail). The students were never engaged in science class, according to their teacher. We were there for several days using the Virus game, and on the second day during lunch two girls came running back from the cafeteria to share some of their theories with us as they tried to catch their breath. The theories were not right, but the point is that they cared enough to be discussing this at lunch.

One teacher shared this comment with us about the Virus game: "I was amazed at the kids' ability to problem-solve...these incredibly low kids who are generally disengaged. I had a look through the post-virus question-naire...without exception...every single kid said that the game was fun and I know from being out in the hallways between classes that the kids were talking to each other about the game, about who is getting everyone sick and so the kids were very engaged. From that perspective I think the game was incredibly successful."

Another teachers stated about the Tit-for-tat game: "But a lot of times you find them getting into what they have to do. And how often do you go into a math class and see a bunch of kids sitting around a table arguing no it's really this, no I swear, look what I did and then the other kids saying no it's this way. And they just kind of like duke it out and sort it out and it's really cool to watch."

It is difficult to get students to feel so personally engaged in learning. While it is entirely possible to create this with virtual media, it is more difficult and is not as good as the real thing. Thus, this principle of *personal engagement* will come back in the design of many of our handheld simulations.

# 7   The Importance of Reality

On the site of a renowned institution, during some routine testing of the groundwater, a team of investigators discovered traces of a dangerous toxin. The toxin, trichloroethylene (TCE), is the same as that implicated in the "cancer clusters" found in nearby Woburn, Massachusetts, the site of the novel and movie *A Civil Action*. Several teams of environmental engineers visit the site where the chemical was found to assess the problem and recommend a course of action. The teams arrive on site and go to work assessing the situation. They interview witnesses and other experts, draw upon local knowledge, research pertinent regulations, and collect samples to document the extent and severity of the problem. All of this needs to be done quickly because the Environmental Protection Agency will send staff to visit the site in the near future and needs to know what, if anything, is being done about the problem.

One team (Team Alpha) presents its recommendation to the leadership. They summarize that they have found the source of the problem, which was dumping of chemicals that happened years ago in the center of the site. The team goes on to say that while the toxin has leaked off the site into the surrounding community, it is not a great danger since the local groundwater is not used as a source of drinking water. The presence of the chemical in drinking water is really the only significant health concern, as its presence in surface water (the local rivers and lakes) is not critical since the toxin is extremely volatile and does not pose a danger there. There are some legal concerns, however. If the toxin leaks beyond the boundaries of the property, the site owners are legally obligated to remediate to make sure that the toxin does not affect neighboring properties. With all of this in mind the team recommends substantial remediation. The way to accomplish this is to position large drill rigs in the middle of the site along the road and to pump the toxin out of the soil. It is likely to a take a few years, but that is what it is going to take to do the job correctly.

In another room a different team (Team Bravo) is presenting its case. They also summarize their case and similarly find that the source of the problem is at the center of the site due to dumping in years gone by. They too have found that the toxin has leaked into the surrounding community and poses no great danger since residents don't drink the groundwater. They concur that the presence of TCE in groundwater is the only real health concern, and that the chemical likely evaporates from the heavily used open bodies of surface water. On the legal implications, they too agree that there are legitimate concerns. But that is where the similarity ends. This team reports it has some concern about the surface water since local rivers and lakes are heavily used for recreation. From experience, however, team members know that these water bodies are already heavily polluted, and this additional toxin is unlikely to contribute anything significant to the problem. Getting back to the legal issues, they feel a measure less drastic than the huge drill rigs might be sufficient. They are afraid that the negative repercussions of putting these drill rigs conspicuously in the center of the site along a highly visible roadway and walkway would be potentially quite damaging to the reputation the site owners. It would create unnecessary panic in the community and strain its already contentious relationship with the site. What would locals think if they see such rigs? It would likely tarnish the reputation of this place for a long time to come. So, Team Bravo instead recommends phytoremediation, which involves planting trees in the most heavily affected areas. The trees will absorb the toxin slowly from the groundwater and diffuse it into the air, which is not problematic. This process will take many times longer than drill rigs, but it will beautify the site, do something about the legal issues, and not cause alarm in the community. Since the health concerns are not relevant here, this is the team's recommended course of action.

How did these two teams come to such different conclusions? The data that they have in hand seems to be identical. They have narrowed the toxin to the same place, assessed the same severity of impact and potential legal and health implications. Yet Team Bravo has factored in additional considerations:

- The water is already known to be polluted, and the additional pollution is unlikely to be a major concern.
- The area in need of remediation is highly visible to cars and pedestrians and would likely draw a lot of negative attention to the site.
- Relationships between the site and the community are strained and should be taken into consideration.

Team Bravo took this additional information and then seemingly weighed the evidence in a more nuanced fashion, and in particular took some of the social/subjective information they collected more seriously than Team Alpha did.

I don't think one can say for certain that one team is more or less correct than the other, but their proposals certainly are different. What would lead two teams to such different conclusions when considering the same data? Perhaps they just have different points of view. I suggest that this is in fact the case, and that they literally had very different views on the data. For one of the teams, figure 7.1 shows what was *literally* their view as they went and collected data.

This team inhabited the virtual world portrayed in figure 7.1, where team members could collect information via virtual sampling devices, obtain virtual interviews from "witnesses," and collect documents to support their case. They could zip across the whole world in a matter of seconds (anything slower was perceived as boring), interact with programmed avatars, and see and converse with other real people represented by other avatars in the world.

The second group had a different view, as figure 7.2 reveals.

This team went out across an actual site. In order to get virtual samples from far-away places, team members had to walk long distances. They could get the same documents and interviews virtually, but they needed to go to the appropriate places to obtain them (for example, for shipping and receiving records they had to go to the shipping and receiving docks). They did not have avators to virtually bump into, but they did have real people to interact with whom they could see.

Based on their responses, can you guess which team had which view? In fact Team Alpha inhabited the virtual world, while Team Bravo went about its activity in the real world. Team Bravo had some additional information that we did not program into the virtual environment. They could see real

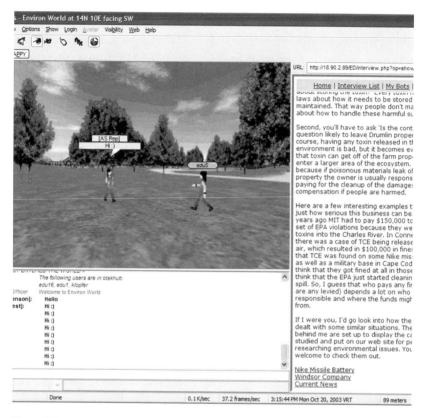

**Figure 7.1**
The virtual version of Environmental Detectives.

people walking by on the street. They could see the high volume of auto traffic that passes by the site every day. They could more readily incorporate their local knowledge of the site into their decision making. This is what likely influenced their vastly different conclusions.

While this comparison is of only two groups, from which we should be careful not to draw any generic conclusions, it does suggest that playing virtual games in real spaces may play an important role in the decision-making process that underlies the learning in these environments. In order to better understand this experience, let's examine just what the players of this scenario saw. Before we examine these cases, it is useful to establish a framework in which both of these technologies exist.

**Figure 7.2**
The augmented reality version of Environmental Detectives.

## Heavy and Light Augmentation

Both the augmented and virtual versions of the game involve technological intervention with which players must interact. Milgram and Kisinho (1994) established a "Mixed Reality" continuum that attempts to define a continuous scale along which environments that mix the real and virtual can be defined. At one end of the spectrum are totally real environments with nothing virtual at all. At the other end of the spectrum are virtual environments

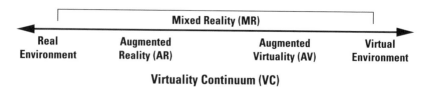

**Figure 7.3**
Milgram and Kishino's (1994) Mixed Reality Spectrum, showing the spectrum of mixes of real and virtual in environments.

with no real components at all. The virtual environment in which we implemented this game would be an example of something close to the latter end of the spectrum. In the middle are augmented reality and augmented virtuality. Augmented reality, in this definition, is a combination of the real and the virtual, which contains more real than virtual. It would be an environment in which one might think about adding virtual elements to a real environment. It could be adding virtual objects or people to a real landscape. The counterpart to augmented reality is augmented virtuality. This may be thought of as adding elements of reality to a virtual environment. Here one might be projecting a real person or an object into an otherwise virtual environment.

This spectrum has come to define what both mixed reality and augmented reality are. But the spectrum is specifically designed to apply to "immersive" technologies, those in which the player is wearing a helmet or goggles and receiving real-time virtual information that is overlaid on the real environment. It necessarily involves a display that is head- or eye-mounted so that it is always in the visual field.

The games that we are creating are designed to use commercial off the shelf handhelds (and a similar case could be made for mobile phones that provide comparable capabilities). As such, these technologies are not immersive, though we hope the experience that they create is, given the elements of the real "immersive" environment that we also appropriate. As such, we need to define a new space for these games. The term augmented reality seems overly constrained in applying it solely in Milgram and Kishino's context. We have used this term more broadly to apply to technologies that combine the real and the virtual in any location-specific way, where both real and virtual information play significant roles. Others have referred to some of these uses as hybrid reality, pervasive gaming, or

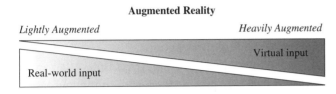

**Augmented Reality**

*Lightly Augmented*                                    *Heavily Augmented*

Virtual input

Real-world input

**Figure 7.4**
A spectrum that defines lightly and heavily augmented environments by the amount of real-world and virtual input given to players.

ubiquitous gaming (Bjork et al. 2002; Magerkurth et al. 2005; Walz 2005; de Souza e Silva and Delacruz 2006). But all of them are augmenting reality with virtual information, so we instead have used a spectrum that defines the weight of the augmentation along a continuum from light to heavy. The weight refers to how much virtual information is provided to the player. A lightly augmented reality has the player using a lot of the physical reality and accessing virtual information quite rarely. Players may look at the virtual information on the order of minutes or even hours. A heavily augmented environment relies on frequent access to virtual information. That information may be accessed on the order of seconds or even continuously.

This puts most of the immersive technologies (helmets and goggles) on the far end of the spectrum as heavily augmented. These environments rely on the fact that you have continuous or near-continuous access to the virtual information. Many of the games that are designed using this technology are ports of purely virtual games in which you obviously also have access to that virtual information continuously. Games such as ARQuake (Piekarski and Thomas 2002) or Human Pacman (Cheok et al. 2004) depend on players being able to respond to the virtual information in a split second. Of course they must also be aware of their real-world surroundings, or they may smash into a wall or fall into a ditch. A lightly augmented world, which is designed around periodic access to virtual information, relies on players making much greater use of the real-world information and accessing the virtual information when certain events happen. Many location-based games (see chapter 4 for examples) fall into this category. Players move about in real space but get information when they move to certain locations or trigger points. While the players might have access to their position in real time, they look at their screens for virtual information only rarely. A virtual environment lies just beyond the right edge of this spectrum, and role-playing games that don't use technology would be just beyond the left edge of this spectrum.

This spectrum is linked to the technology employed. Displays that are in the players' field of view are likely to provide them with continuous input, but a screen on a handheld or mobile phone is likely to be accessed much less frequently. This need not be the case, however. One could design a handheld game that requires the player to look at the screen quite frequently, or a head-mounted display that provides only event-based information to a player.

The game Savannah (Facer et al. 2004), designed by NESTA Futurelab, is an educational augmented reality game. It provides a heavily augmented reality via handhelds to students. Students in this game are role playing as lions hunting on the savannah. They need to track their prey and stay away from potential hazards such as hunters or other lions. Players are given headphones that provide them with continual audio cues, and a handheld on which they can track their location. There are events that happen periodically (they catch prey or get pounced on by another lion), but they must continually track their location relative to others and pick up on cues like scents on a near-real-time basis.

The heavy to light augmented reality spectrum is also correlated with the kinds of games one might create using these different technologies. Typically one might associate games that are not connected to the real space to be heavily augmented (using the real world as game board), while games that are tightly linked to particular places to be lightly augmented. This connection will be discussed in the next chapter.

**Environmental Detectives**

The two chemical disaster scenarios described earlier in this chapter were both part of a game that we created called Environmental Detectives (ED). ED was designed as a part of the Games to Teach project at MIT, funded by the Microsoft iCampus initiative. This particular game was the first of what was to become a long series of augmented reality handheld games (ARHGs). These games fall on the light side of the augmented reality spectrum, providing players with some occasional virtual information that they must integrate with their real-world experience. We chose a lightly augmented approach for this first game. The development of ED as an example of a lightly augmented reality game was motivated by a desire to create games that could address important disciplinary practices in realistic ways. We thought that we could create some compelling authentic scenarios with the corresponding strongest learning outcomes by setting these games in real space instead of the common tactic of setting games in virtual spaces.

The time (a. 2001) was right for such an initiative, for handheld computers were then rapidly increasing in their capabilities and had become the cornerstone of such an activity using commercial, off-the-shelf hardware. That applicability to commercial off-the-shelf products was important, as we did not want this to be a purely academic initiative in which we would be tied up in proprietary hardware that would be of little use outside a research project. Fortunately, handhelds were offering several new properties that delivered unique affordances (Klopfer, Squire, and Jenkins 2002):

• Portability can be transported to different sites and move around within a location.

• Social interactivity can facilitate exchange of data and collaboration with other people face to face.

• Context sensitivity can gather data unique to the current location, environment, and time, including both real and simulated data.

• Connectivity can connect handhelds to data-collection devices, other handhelds, and to a common network that creates a true shared environment.

• Individuality can provide unique scaffolding that is customized to the individual's path of investigation.

Together these features could be combined to create a location-based simulation game that we felt could be both authentic and engaging. Given that we were at an institution where engineering is a significant academic focus (albeit taught primarily traditionally), we targeted engineering as our first domain. In consultation with environmental engineers we identified a particular problem with environmental engineers-in-training (Nepf 2002). New engineers often struggled with navigating primary (data they collected themselves through scientific measurement) and secondary (desktop research or interviews with witnesses and experts) information when conducting investigations. While the faculty tried to teach this skill in the classroom, it was hard to do as it really only came along with practice. This seemed like the perfect skill to try to teach through a game situated in the real world. We could engage students in the study of this disciplinary practice through authentic technology-supported investigations.

Students playing the game were given a briefing containing information similar to that described at the beginning of this chapter. They were told that routine groundwater testing during the construction of one of the buildings on campus showed evidence of a chemical in the groundwater. They were being brought to campus as a team of environmental engineers to try to determine where this chemical had come from, how far it had gotten, what the legal and health ramifications of its presence were, and what the university should do about it. The problem needed to be dealt with quickly because the Environmental Protection Agency (EPA) enforcers were coming to campus for their annual visit soon, and the university had previously been in trouble with the EPA and therefore wanted to play this one by the book.

Each pair of students was given a handheld computer with GPS that would perform a number of functions. Students' main view was a schematic map of the area (figure 7.5a). On that map they could see their current location marked by an icon. As they moved around on the site the icon moved with them. Additional icons indicated the location of virtual experts and samples that they had already taken.

The primary tools that the student teams had at their disposal were:

Virtual interviews   Experts and witnesses were scattered around the campus, positioned in locations that were linked to the information that they provided. The president could be interviewed near the president's office. Experts on geohydrology were near that department's location. Witnesses who may have seen dumping long ago were in offices that were near the dump sites. Interviews always contained text and often contained additional images or video.

Electronic documents   Documents could be obtained either directly from sites or handed to teams by the witnesses and experts in the game. This background information was taken directly from actual documents containing real information.

Virtual samples   Everyone was given three drill rigs that they could have out at any one time. The drill rigs could provide data on the level of the

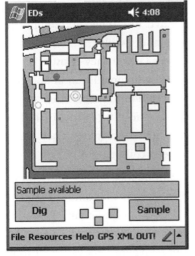

(a)

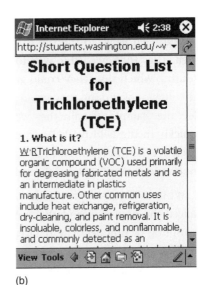

(b)

(c)

**Figure 7.5a–c**

Screen shots of the first implementation of Environmental Detectives, showing (a) the schematic map of the area; (b) representative documents; and (c) on-location video.

toxin at any given point on the campus. To use the drill rigs the player would need to go to the site that they wanted to sample, and "dig" the well. They then needed to wait for several minutes while the well was dug (during which time they could go on to other tasks if they chose). When the digging was done they could go back to the original location and collect the "sample." Finally they were given an option to process the sample quickly in the field (with low accuracy) or to send it to a virtual lab that took several more minutes to process (with higher accuracy). They could not reuse the drill rig until the sample had been collected.

While others had previously created location-based games, ours introduced the notion of underlying data and models in the form of different sampling techniques. The data was provided by an underlying model of the spread of the toxin through space. *Combining quantitative and qualitative information* turned out to be a key design principle for later games. It took what was a platform for location-based Web browsing or place-based narrative to a new direction in which virtual data was layered throughout space. The kinds of games that this enabled provided fertile ground for learning about data collection, analysis, and sense-making, which are important twenty-first-century skills.

Environmental Detectives was designed to provide some of the realistic challenges of conducting environmental investigations. Most students wanted to simply put the drill rigs down everywhere to get samples, but this was not realistic. Instead they needed to spend time planning where they wanted to put the rigs, and use them judiciously. It took time to sample and time to get to the sampling locations. All of this timing needed to be factored into the students' decision making, for which there was a great deal of variability among the teams. Adding to the authenticity was the fact that there was no perfect solution to this problem. Some solutions were better or worse, but none were perfect. In terms of FITness skills, this was *managing problems in faulty solutions*.

Additionally, students needed to deal with real limitations in the environment. Some students playing a similar game at another location discussed some of the barriers to their plan of attack:

Stacey:   There's a fence there. I can't get over it.

Gina:   Then I don't know what we're going to do. We're stumped. Let's call the guy [facilitator] so we can find out what we're doing.

Stacey:   What does it look like?

Gina:   We're close. That's the thing.

Stacey:   Okay, fine. Can we go over this [barbed wire] fence?

Gina:   I don't know.

Stacey:   Maybe we can get on the other side by walking somewhere else.

Louis:   Maybe we can walk the fence. No, there are trees.

The real space provided real barriers and caused the students to make some decisions of the kind not faced in a virtual world where the consequences are merely virtual. Climbing a barbed wire fence in a virtual world is either possible or not possible. Climbing a barbed wire fence in the real world has well-known consequences.

Conversation focused on the game is important, but the time spent walking from one location to another, and observing one's surroundings in the process, leads to another kind of conversation. It is that more casual conversation, related to the game and the surroundings, which likely influenced the players in making different decisions and weighing different factors across the virtual and augmented worlds. For example:

Bill:   [*Walking near the river*] We know that it's in the Charles, which is already disgusting. It's possible that TCE is such a ridiculously small affect compared to the big mess of the Charles, and I have friends by the way who study the Charles River and are not impressed. So, that's a possibility. We also know that the water isn't used for drinking.

Jenny:   We used to go canoeing on the Charles River. And we always had to watch out. People fell out of their canoe, their eyes were stinging and stuff.

These frequent conversations about the community and their surroundings were important. They elicited responses about the players' past interactions with the place in which they were, and caused them to incorporate information about their surroundings—pedestrians passing by, information about other buildings, stories that they had heard, and so on.

Sometimes the way in which that information was ultimately incorporated into the story by the players perplexed the passersby. Occasionally players would ask other people wandering by for information pertinent to the game. As these games were *near* reality, that information could sometimes help them. For example, someone might ask for the location of a particular building, where the university president's office was, or if a passerby knew anything about the local groundwater. All of that information would be relevant to the game. However, sometimes players struggled with where the game ended and reality began, asking befuddled passersby about fictional aspects of the game. This was greeted with indifference, confusion, or alarm. In this case, where Environmental Detectives was being played at a nature center, the person who gets caught up in the story has cause for alarm:

Stacey:  We're trying to find if there are any toxins here. Do you know of any toxins?

Visitor:  Toxins. I don't know of any toxins.

Gina:  It is in the game. I *think* it is all in the game.

While the two girls likely understand that there is no imminent danger, and they are not empowered to take action on any such toxins, this interaction demonstrates that they are invested enough in this investigation to think that perhaps a passerby will have relevant information. It feels authentic enough to them that they have lost some sight of where the game ends and reality begins.

## Nothing Beats Reality

When asked to compare the Multi-User Virtual Environment (MUVE) experience with the augmented reality experience, students gave us a wide range of responses. Some students liked certain aspects of one environment or the other. Those students who described themselves as "kinesthetic" learners, or "tactile" learners in general strongly preferred the augmented reality simulation. "I learn better by doing something myself," one student noted, "so I will definitely remember the information we encountered in these experiments better than if I had just read about the process."

Another player felt more comfortable in the augmented reality simulation, explaining, "I liked the freedom of being outside. I liked being able to physically touch the map and I liked being able to really look around for clues about where we were. I did not feel disoriented in the augmented reality simulation. Whereas I did feel disoriented in the virtual reality simulation."

While some people (myself included) sometimes feel "directionally challenged" in the real world, we are better suited to picking up the cues and appropriating our skills at navigation in a real-world scenario than we are in an avatar-based virtual one. We are able to draw upon what little skills and techniques we have developed through life in navigating these spaces. Yet, other students enjoyed being able to get a very rapid sense of managing the situation in a more expeditious manner. Through the virtual environment they could zip around and collect the data that they needed which was expedited without the physical constraints of the real world. One player offered the following explanation for their enjoyment of the virtual environment: "I think what I enjoyed the most was actually being able to get the answer. I was concerned at the beginning that we would not get the answer but somehow as you navigate your way through the world, interview different people (collect data really), the challenge becomes manageable. I think this is a good tool for learning."

I argue that getting "the answer" should not be the goal of this activity. It is really about the journey rather than the destination. In real situations such as this, we may never know "the answer," but rather collect as much information as we can to make a well-informed decision—such as happens in many events in real life. This skill of being able to make decisions with information that is less than complete, or managing faulty solutions, is a critical twenty-first-century skill, but one we don't teach very well in most classes where answers and problem sets are graded right and wrong. In practice, getting "the answer" is either impractical or undesirable due to great costs (time and money). A student reflected on this difference with the AR simulation, in which it was harder to get the answer due to the intricacies of the real world that constrained their investigation: "The AR simulation really appealed to me because it felt like an activity. Like we were going somewhere and in doing so, I think that I learned more about the topic and I felt more like a person in the field. Although we didn't complete the exercise...in fact, BECAUSE we didn't complete the exercise, there was a lot for us to think about and the places and interviews that we didn't encounter left us thinking about how they would have fit into the final conclusion." The student added that "the VR simulation was interesting because it was somewhat easier to see an outcome at the end. The data could be presented with easier access and I really didn't have to work to get it. In this respect the VR simulation seemed a bit more like an executive position...directing researchers around the field."

There was some consensus in class discussion that the virtual (MUVE) version gave the players a better sense of planning the investigation, but that the augmented reality version gave a better sense of actually conducting the investigation. Said one student: "Although I felt that the process went fairly smoothly, virtual reality cannot mimic real-life conversations and design of an environment. However, I did think that it provided a thoughtful way to explore a problem and think about its causes and consequences and this does not necessarily have to take place outside....Then augmented reality is a great way to simulate, more so than VR, a real-life experience. Being in the field enables you to get a much better sense of the terrain that you are working with, and it allows for a more authentic feel."

The problem is that the "executive" position is how we train students to typically view these problems. However, actual problems don't stand up well under the executive model. Real problem solving involves a great deal of trade-off and assessment of when the information is good enough. In fact, I argue that the *hard* part of this problem solving is that assessment and therefore that is what one should design these learning games around. The real world is messy and imprecise. Games are an excellent way of capturing the decision-making process that one most employs to navigate this incomplete picture. However, virtual environments that are purely digital can do precisely the opposite and convey that everything is exact and knowable to many digits of precision. Even when that data has variability it may be perceived as accurate to users who interact with that environment. The augmented data comes embedded in the real world, which is known to be irregular and imprecise. Situating augmented reality games in that world gives players the expectations that the game incorporates the "unknowability" of the real world.

This lesson was indeed learned by some of the ED players. In an interview after the game one of the players was asked what the most important thing was that she learned from the game. She responded, "That it's difficult (and time consuming) to be able to pinpoint the source of problems, but that it's important to keep going until I can come up with a reasonable explanation." This kind of learning, this meta-reflection on problem solving, is what we strive to create and often only hope we can attain. In this case, the student was able to define a goal of "reasonable explanation" instead of "the answer," a much more authentic goal.

What is interesting is that even when students preferred the virtual environment, they gave a nod to the influence of the real environment on their learning and problem solving. One student stated that the virtual environment "seemed very far removed from reality—[I] didn't care as much about the problem because the contaminated water was just on the screen."

This indicates that in games such as this, when the outcomes have complex rationales involving scientific as well as social, economic, legal, and public policy implications, that the augmented reality version can capture many of those real consequences, while maintaining simple design, in a way that can't be touched by virtual environments. In other words *nothing can provide a sense of reality like reality*.

## The Authenticity of Roles

While I cover both authenticity (chapters 8 and 10) and roles (chapter 9) later, it is interesting to note the differences in how the players felt about their roles in the virtual and augmented environments. Authenticity has deep connections with learning and motivation. Learning must be thought of as an interaction between the learner and their environment, including both the physical environment and the other people in that environment with whom they interact (Lave and Wenger 1991). Brown, Collins, and Duguid (1989) argue that transferable learning cannot take place outside of an authentic context. It is only through communities of practice around authentic problems that learners can begin to acquire understanding.

Shaffer and Resnick (1999) describe four different views on what authenticity means in learning activities: (a) materials and activities aligned with the world outside the class room; (b) assessment aligned with (what students really should learn from) instruction; (c) topics of study aligned with what learners want to know; and (d) methods of inquiry aligned with the essential practices of the discipline.

They further describe the notion of "thick" authenticity, which combines all four of these elements, and argue that an activity can really only be authentic when it combines *all* four.

It seems clear that there are degrees of authenticity that vary based on the degree to which they match these four elements. While I later discuss what defines authenticity, what is clear is that to at least some extent, *authenticity is in the eye of the beholder* (Barab, Squire, and Dueber 2000).

More specifically, authenticities lie in the learner-perceived relations between the practices they are carrying out and the use of these practices

(Barab, Squire, and Dueber 2000, p. 38). The benefits of the authenticity of an experience may only matter if the person involved in the activity becomes aware that the experience truly feels authentic. We as the designers of Environmental Detectives created everything to be nearly identical in these two environments. The dialogue from the characters, tools that the players could use, and all of the information to which they had access were the same. However, the interactions between the player, the activity, and the environment is uniquely determined by how the player sees themselves in the context of the activity and the environment, and thus these "identical" exercises may convey radically different authenticity to the player. Some of this difference may be attributed to the intangibles that come along for free in the real world—the pedestrians walking by, the sounds and smells in the environment, or memories associated with particular locations that are evoked when people experience them.

But this may also be from a physical and mental feeling of doing the same activities that a real person might in these situations. Thus augmented reality goes further in capturing what Dewey (1938) might classify as an *experience*. While actual real-world activities—for example testing the levels of chemicals in a body of water—may also provide a great sense of authenticity, they cannot practically teach about certain topics. Teaching about environmental engineering needs to involve more than taking samples from surface water; it should involve the practices, tools, techniques, and thinking of environmental engineers, or the best approximation that we can get.

We might compare the use of environmental data taken from PDAs in this activity to similar activities using probeware (Tinker and Kracjik 2001). Using handheld probes attached to PDAs students can collect data about the soil, water, and air. This data (e.g., salinity or dissolved oxygen in water, pH in soil, or ozone in the air) is much like the data environmental engineers must collect. In that way, these activities are authentic. However, they are not typically set in the context of a real *problem* to be solved. It is not a part of environmental engineering practice to just go around and collect data. Instead, such practice is conducted in the cause of trying to solve a problem. This "method of inquiry" creates a radically different activity, one in which the participants must collect, evaluate, and apply information qualitatively differently.

In contrast, the augmented reality experience helps players feel like environmental engineers. One element of this is the planning for real space and real environments. As one ED player noted, the augmented reality environment "created a great sense of trying to solve a 'real' problem. Because we had to physically walk from one place to the other, it was critical to make a plan and revisit our plan during the process."

The simple act of walking from one place to another, a process that is very hard to accurately simulate in a virtual environment without becoming boring, gave this player a sense that this was a "real problem" that needed to be solved. So one sense of authenticity was *planning that required taking reality into account.*

Players also reflected on what it was that made them feel as if their role of environmental engineer were authentic. Designers of virtual worlds often attempt to convey that sense of authenticity by connecting the player to the avatar. They let the player dress the avatar up in the role as they see appropriate. Or they make the avatar look like the player so that the player can see themselves inside of the game. They become *embodied* in the game itself.

Virtual worlds go to great lengths to allow this customization to create a deeper connection between the avatar on the screen and the learner behind the controls of the game. This connection is much easier to create in an augmented reality environment, though sometimes that connection is misunderstood. I was a member of a panel at a conference recently that featured presentations about both augmented reality and on-screen virtual worlds. The discussant challenged all of us panelists about the degree to which we allowed customization of the personas that people portrayed in our game environments. The virtual world panelist described the "avatar designer" that they had in their game, which permitted changing the avatar's clothes to match that of the player or the role that they were portraying. Players could even let their male character dress as a female if they wanted. When it came to describing similar features in our augmented world, we said that we get the clothes that exactly match those of the player for free. If you want to match the look of the role, there is a second-hand costume shop not far from campus where players could go. And yes, if they wanted to, we'd let a male player dress as a female if they chose to come to class that way that day.

The fact is that environmental engineers in our game look like "regular people." The important thing is to see an environmental engineer who resembles yourself, not necessarily making yourself look like some stereotypically crafted professional. The players in the augmented reality game used the experience to feel like they were environmental engineers (or environmental *detectives* as they sometimes referred to the role, as the name of the game conveyed).

One player tried to articulate the elements of the real world incorporated in the augmented reality game that made them feel as if it were authentic: "It was really cool going to the construction site and taking a toxic sample and then moving over to different buildings and conducting interviews with various people on campus. I like the physical distances traveled to 'gather' the data. I truly felt more like a detective in this environment—the technology aided in this feeling."

Another player felt more absorbed in the augmented reality world, commenting that "the augmented reality situation felt authentic to me (when everything was in working order). The MUVE, however, felt like a game and as a result I had a hard time taking it seriously. I found it much easier to be 'absorbed' by the augmented reality simulation—I relished playing the role of an environmental scientist."

This player also referred to the connection with decision making and taking the game "seriously." Players take the decisions seriously, which influences what they ultimately might decide to do in this situation. They cannot arbitrarily weigh factors and place judgment. They need to take into account all of the information that they have and decide accordingly.

In some instances the jump from the player's real role in life to the role portrayed in the game was too big a leap. In these instances, the players might take intermediate steps toward feeling the authenticity of the role. That is, they feel like the situation is authentically based on some more proximate role. One student felt "like a cutting-edge MIT student running around with the handhelds—I felt like I belonged there. The GPS / map overlay was cool too. Again the task was rooted in understanding." While this comment would seem strange coming from an MIT student, in fact it came from a student from another nearby university, typically less associated with that techno feel.

While this example is meant somewhat in jest, it does suggest that these games can create small steps in perspective taking, if not giant leaps into the actual roles they are meant to convey. We may not in fact expect students to take giant leaps in these single instances of the game, but they may be coached along a progression of such experiences over time to really see themselves in these roles. Nonetheless, there is likely to be learning from even these small steps.

## Face-to-Face Communication

One final note on the differences between the real and the virtual implementations of ED has to do with forms of communication. With the rise in capabilities of online meeting systems, and the coincident rise in hassle associated with airline travel, the airlines have started to run commercials emphasizing the value of face-to-face visits with friends, family, and clients. Face-to-face visits can certainly convey the value of relationships, but they can convey much more as well.

Studies (Klemmer, Hartman, and Takayama 2006) show that less-constrained styles of interaction can not only enhance communication, but also are associated with better *thinking* as well. The social interaction, both within groups and between groups, is a critical component of these activities, both on line and off. Allowing people to freely communicate through whatever means they choose (gestures, tone of voice, facial expressions) influences the interactions between the players, and further may affect their thought processes during the game. Methods of virtual communication are improving through more realistic avatars, video chat, and enhanced audio. At the same time people are adapting to these new modes of communication with their own forms of expression (e.g., emoticons ;-)), but it will be quite some time before we can fully capture the experience of meeting face to face.

# 8   Location Matters: The Role of Place

When discussing the value of handhelds in education one day, someone told me that he saw two great uses for them, "those in which location doesn't matter and those in which it does." While it is easy to see the world in such dichotomies, this division makes more sense than it appears to at first glance.

Mobile devices are a good fit for applications that can take place anywhere at any time—that is, those applications for which location doesn't matter. One might want to check one's email, jot down a thought, communicate with a buddy, or see the latest news headlines. The uses for handhelds are ubiquitous and not tied to any particular location. They can happen on a train, waiting in the doctor's office, on the way to a meeting, or as the case often seems to be, while in a meeting. These applications are sometimes classified as *productivity* applications, and are content-independent. They are of generic utility and are not tied to any particular subject matter.

On the other hand we have applications for which location does matter. These applications might give you the time that the next train will arrive at the station you're at, tell you restaurants that you might enjoy in the neighborhood, or tell you the history of that same neighborhood. For these applications location is critical. It is much less relevant to know what restaurants you might like that are not nearby, or the train schedules of some other train line. The user's current location is one of the key inputs to these applications. The other key input is some category of activity with the system—transportation, dining, or tourism in these examples. That makes these applications content-specific. While some sophisticated systems may actually deduce what it is that you want to do by your location (e.g., if you are walking near the train at the end of the work day then you probably

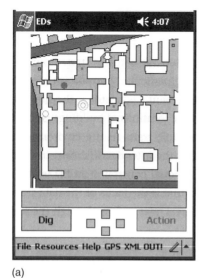

(a)

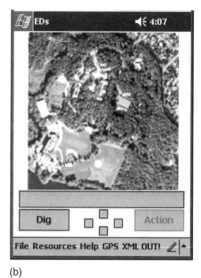

(b)

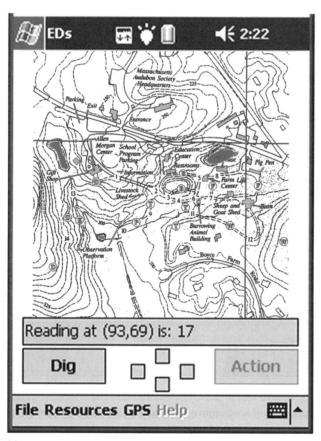

(c)

and groundwater, so we needed to have a space that was of sufficient size to contain the spill.

Once we got a sense of the space, its assets, and landscape, we could reposition all of the interviews and information from the university-based version of ED to the center's location. The university president's office became the Audubon Society headquarters. The site near the Charles River became a site near a pond. Several elements had direct parallels, such as the library and facilities buildings, which made porting those trivial. We also had to grab the GPS coordinates of three of the four corners of the map to input into the system, and scale this to the contour map that was chosen by the center staff as a good representation of the area. The contour map was particularly relevant because the site contained places that were a part of two different watersheds. Land on one side had water flow into one watershed, while land on the other side caused water to flow elsewhere. This meant that the location of the spill of the toxin was critical and could send it to two entirely different places. Certainly here, even as far as just the map was concerned, location mattered.

We went through a similar process at the second site, the school campus. We chose a space of comparable scale and identified buildings that were similar in nature that could be functionally mapped to buildings in the other two scenarios. Again, even as far as the map was concerned, location mattered. We didn't just choose arbitrary points to represent the different buildings, but systematically found buildings that made sense in the context in which we were using them. The school happened to have a great aerial map of the site that we could use for these purposes, though it had to be scaled and rotated to fit with the GPS coordinates that were measured at the site. What resulted from these transformations were three maps as shown in figure 8.2, with the first being the original university-based game, the second at the nature center, and the third from the school campus.

The students demonstrated that the locations also mattered to them. As they planned their routes and their investigations, they had to take into account many of the factors that had been anticipated in the design—how

◄ Figure 8.2a–c
Maps of Environmental Detectives at different locations. Each of the maps emphasizes different aspects, whether it is (a) the layout of buildings; (b) the topography; or (c) the foliage and landscape.

to get from one point to another, how the information from one point might help them in determining where to go next, and so on. The location of information was perhaps the single most influential factor in students' planning of the investigation.

Here one group of high school students decides where to go in their investigation of the site:

Stacey:   Did they tell us to go to a specific place? Like what is the Audubon thing? So if I can find it on the map [meaning the paper map] then I can find the box near it.

Gina:   It says you want to go to the toxicologist and you can find information at headquarters.

Stacey:   Audubon Headquarters?

Gina:   Yeah.

Stacey:   Where is that?

Gina:   How they dealt with it, ... Where are we going now? There is another close one [interview].

Stacey:   Let's go to that one [pointing to the learning center]. We just traipsed through a field.

Louis:   I like how he [the character in the video] was standing up there [pointing towards the house] and reading it.

Gina:   Yeah, I know.

Louis:   He got to stand at the house and we had to stand in the water [in the field].

Stacey:   I know. I am so wet.

Louis:   My socks are so wet.

This group (notably a high school group on a field trip to the nature center) takes strong cues from location as to what they should do next. They'd like to obtain a nearby interview (using proximity as a cue) and they'd like to not get their feet wet (using features of the actual landscape as a cue). While these may not be used in the ideal way that we as designers intended, they are certainly highly influential in the decision making of the students playing the game.

## Relocating Content

The connection of the interviews to local buildings made the space easier to navigate. When a player saw the real sign that said Library, they could use that information to predict what kinds of resources that place might hold. It also influenced the way the players navigated the game space. But that alone was not sufficient for ED to make sense in a new space. The original backstory concerned the discovery of chemicals in the groundwater during the construction of a building on campus. It challenged the players to deal with the legal and health ramifications for that location, which took into account things like where the groundwater went, the size of the campus, and the presence of surface water (a river) that was used by the public. The use of the chemical in question also needed to make sense for that location. Simply placing that identical story in one of these other locations would have made it seem out of context.

Relocating this game to a new location involved rewriting the narrative to make sense in it. It meant thinking about what parallels could be made between the places, and also involved using the unique features of those new places to attach the game better to the new location.

The nature center turned out to have two unique features. First, it was a working farm. The livestock on the farm drank water that came from a pond in the middle of the nature center. We used this as a replacement for the groundwater on the university campus. The second feature was that it used to be a NIKE missile base (before it was a nature center). It turned out that it was only a command and control center. The actual missiles were located elsewhere. But that provided some rich and interesting history for the location. It also meant that we had a real red herring that we could build into the game. The introductory story for the nature center then went as follows:

Hello. My name is Pete Nomenson. I am the one who hired you to research the environmental issues. Thank you for coming to Drumlin Farm today. We've brought you here to do a standard environmental investigation of this site before the state purchases it to turn it into a working historical farm.

Normally, this would be a routine job. I wouldn't expect to find anything. But the local veterinarian, Karen Richmond, tells me that she's seen a lot of sick cows and pigs coming from this farm—more than you would normally expect. This site was a NIKE missile base until the 1970s, so there were all kinds of crazy chemicals and stuff stored here. Heck, it could be anything.

My hunch is that the source of the animal illness is something in the groundwater or maybe the soil. I don't trust those underground silos one bit. Around the farm are permanent wells and bore tubes used to sample the water systems. I'm not so sure how they work; you should talk with one of the scientists over in the Hathaway Environmental Library. Or, you can just head out there and start collecting samples.

I want you to prepare a report detailing the current status of the farm. I want to know: What, if anything is causing these deaths in farm animals. Is this site safe to turn into a farm that is open to the public? If there is anything wrong, what caused it, and how hard will it be to clean up?

Fortunately, there are plenty of scientists on site who know a lot about the history of this site, chemistry, and animal husbandry. Please keep in mind that our main goal here is to *"determine if this land is worth buying."*

The school site also had some interesting features. There were very old buildings that had mysterious histories. There was varied use of water at the site—groundwater for some purposes and public water supply for others. School staff also had some incidents in the past with illegal dumping on the edge of their campus that could also be incorporated into the ED story delivered by the headmaster. The backstory for this site then became:

Thanks for your quick response. I am glad that you could help us with this important environmental issue. We need you to conduct a thorough investigation on the Nobles campus.

Normally, this would be a routine job. I wouldn't expect to find anything. But the water chemist at the water treatment plant across the road from Nobles tells me that he's seen some surprising results on some routine water tests—higher levels than you would normally expect. We sometimes hire contractors who use different types of chemicals on campus, and I suppose it's possible that these chemicals may have somehow affected the water supply. Fertilizer, machine oil, paint, cleaning fluids. . . . Really, it could be anything.

And remember, we're talking about well water here. Nobles gets its drinking water from the town water supply, and that's tested constantly to ensure that it's safe to drink. But something is wrong with our well water, which we use for the sprinkler systems and the swimming pool, and that's showing up in these tests that the water chemist ran.

My hunch is that it's something in the groundwater or maybe the soil or something. This was once a farming community a hundred years ago. There are a number of private wells in some backyards around here, and of course the Dedham-Westwood Water District treatment plant is just across Bridge Street.

It would be a good idea for you to meet with Leslie Chen, the woman who oversaw the digging of the wells across the street at the water treatment plant. I believe she's at Baker [Hall] right now and can tell you how to obtain the well samples you're likely to need. Or, you can just head out there and start collecting samples right away.

I want you to prepare a report detailing the current status of the campus. I want to know: What, if anything, is causing these high readings? Is the campus ground water supply safe? If there is anything wrong, what caused it, and how hard will it be to clean up?

The format of the report will be a written assessment of the points above, and you should be prepared to participate in a class discussion of your findings and analysis.

Fortunately, there are plenty of scientists and faculty and staff members on site who know a lot about the history of Nobles, chemistry, and the physical layout of the property. Please keep in mind that our main goal here is to "determine what happened, and why."

Each of these games was *about* the location, not just held at the location. To further attach the game to the place, we shot pictures and video on location so that players felt the characters in the game were really there. In most cases we used the names of real people who worked in those roles at those sites. In one video I role played someone who worked with the farm machinery. That video was shot in the location that students would be seeing the video (figure 8.3). Players often remarked on these photos and videos, talking about "the guy from the place we saw earlier" or other references to people actually being in particular locations.

**Transporting to New Users**

Each of the audiences that ED was designed for had different experiences and learning objectives. Taking this into account, we redesigned

**Figure 8.3**
A video sample from Environmental Detectives shot on location at the site. The background shows some of the farm machinery used in the story.

documents to be shorter and more digestible for younger students. Similarly, for this age group interviews were often more concise, only telling the most relevant details. This worked very well for the students playing this game on their own campuses. Both the university students and the private school students readily engaged in problem solving and wanted to find the best solution to the problem, even when finding that solution turned out to be too complex a task. These findings have been echoed in other augmented reality game implementations. In other games with connections to the environment, Squire and Jan (2007) found that students who knew an area well successfully applied preexisting knowledge and feelings about a locale to a complex problem, and in turn transferred that understanding back to real-world issues.

The students at the nature center had an entirely different experience. If we have one mantra in the design of our location-based games it is, "It is not a scavenger hunt." The key to a successful augmented reality game is challenging the players to solve complex problems using the combination of real and virtual information. While collecting information at the sites is critical to solving such problems, it is the connections between those pieces of information that make this a powerful learning experience. Making those connections, however, may be a function of how much the players know or care about the space in which they are playing. The students at the nature center were on a field trip to a place they had only learned about remotely. It was not their community, and they felt no particular connection to the landscape or to the place. In many ways, it was somewhat of a hybrid between the real and virtual experiences of the previous chapter. It was a real space in which they saw a real landscape, people, and animals, but they were disconnected from that reality by being dropped into that location for a few fleeting hours.

As a result, for many of the students in this scenario, the game became a scavenger hunt (Klopfer and Squire 2007). Rather than seeking connections between the pieces of information in order to solve a problem, it was just about hoarding that information. It was about "collecting the dots," instead of "connecting the dots," which makes seeing the complete picture impossible. When two teams encountered each other in this game they compared notes not on the content of the information that they had found, but the quantity of information.

Karen (team 2):   How many [interviews] did you get so far?

Louis:   None, nothing.

Stacey:   We've only gotten one box. How many have you got?

Karen (team 2):   One so far. We were going for another one.

Bill (team 2):   Three. Oh. You meant the boxes?

Gina:   Did you dig?

Bill (team 2):   Yeah.

Gina:   Can you dig anywhere?

Bill (team 2):   Yeah. I think so—I did.

Gina:   Cool. We got an interview. That's all we did. We don't have much time. We have to go.

This interaction had the feeling of a scavenger hunt not only because the students were merely collecting boxes, but also because they were competitive about the number of boxes that they have. These teams could have helped each other by sharing information, but instead they only compared what percentage of the task they had completed. Later the team assessed its success in this activity, reflecting again on the number of boxes that they had:

Gina:   I am so happy that we have at least one box.

Louis:   Yeah.

Gina:   And we have that it is the TCE chemical. That is what they think it is, so we have something to say. I am quite happy about that.

Here they do reflect on the content of that box, but never engage in processing this information by connecting it to the place or the data samples they collected.

When considering why these students were not able to engage in the "real" part of this augmented reality, we can look to additional parts of their background and behavior. While the dialogue was tailored to their level, and they were briefed on the scenario beforehand, this class had little culture of engaging in "hard problems" and likely needed additional scaffolding to effectively take on this task:

Cynthia:   How are we supposed to make recommendations?

Ling:   I don't know.

Ethan:   Just read off of the information that we got.

Cynthia:   I thought we could dilly dally but we actually did work.

Ethan:   For once.

Additionally, there was a culture of "efficient learning." This was an Advanced Placement class in which there is a lot of material to cover in a relatively short period of time. Under such circumstances most approaches favor "efficient learning," packing as much content as one can into the year. This in turn means a lot of lectures or activities in which the information is packaged for students in an easy-to-understand way.

In reflecting on the activity afterward, one of the students suggested that the game was not a particularly efficient way of conducting the activity.

Nick:  We didn't get to read everything, because we were just going (*snaps three times*)—boom, boom, boom—running and getting chased by a guy with a knife . . . well, it was metaphorical knife. Maybe we could have all of the people in one room and talk to them all like around different places in the room.

Their teacher asked if Nick thought that running this all in one room would have been better than the outdoor experience. The student further emphasized the importance of efficiency in conducting such learning activities, but then gave a pointer to the fact that there may be value to the game taking place outside:

Nick:  It would be more efficient, but maybe the point of it is to go out and walk around and see everything too. I don't know what the objective is, but if the objective is to get all the info real quick, then the best way is to do it here [in one room].

This team expressed that they didn't know what the purpose of the outdoor portion was, and that if they were simply expected to learn the information it would have been more efficient to give it to them. This failure to put the different pieces together—the physical environment, along with the virtual information—seems to have contributed to this team's failure to make sense of the situation. This confusion was not universal among the students in this class. A student on another team suggested that she had used some information gleaned from the outside in her assessment of the situation.

Maya:  The way the water traveled? If we were up on the hill and the water would go down . . . so we thought if it was the water contaminating down.

There are several lessons to glean from the design and implementation of these games across locations. First, and most generally, this case demonstrates that it is hard to undo the classroom culture of efficiency. With many demands on teachers, students, and schools to perform on vast, comprehensive standardized exams, many educators have chosen the path focusing on the most efficient instruction in their classes. Many existing classroom technologies are designed to support this usage. Students use the technologies for rapid access to information, an important and necessary skill, but one that is insufficient in the current world. That information needs to be contextualized and processed. That part takes time, and practice. Jumping into complex problem-solving spaces may work well for students who already have at least some skills in defining problems and performing open-ended investigations. But students without those skills may just flounder. This is as an indication that future environments should be designed with additional scaffolds to help students along in their investigations. While such support may be as straightforward as contextual help or intelligent agents, it could also come in the form of a more dynamic environment that could, through the events of the game, guide students along a path or multitude of pathways and provide them with more direction when it is needed and less when it is not. Some of these features are incorporated in games discussed in later chapters.

With regard to augmented reality games, the real world plays a more significant role in the game if the specific place already has meaning to the players. If we think of the place as a key element of the narrative, the physical attributes of that space alone can provide some meaning to the players in terms of what they can observe on site. Players might be able to obtain meaning from looking at the natural or manmade landscape, observing people or animals moving about, or being able to ascertain certain properties based on the community. However, if the players have previous knowledge or experience with that place, all of that meaning is brought to the game-play experience. This may convey deeper meaning for the place, while it also may bias the players in certain ways. It may be possible to design games that rapidly create a connection between the game, the player, and the place, but previously existing connections are key elements to design with or around (in the case of bias).

## The Localization Spectrum

The heavy and light augmentation spectrum (see chapter 7) provided one way in which to look at augmented reality games in terms of the relative amounts of real and virtual information that were used by players in the games. Another, and complementary, way to look at these games is by

**Augmented Reality Localization**

*Lightly Localized*                                        *Heavily Localized*

**Figure 8.4**
The spectrum of lightly to heavily localized games corresponds with the portability and specificity of those games.

how tightly the games are connected to a particular place. This localization spectrum defines the portability and specificity of the game along a continuum from lightly localized (highly portable and quite general) to heavily localized (of limited portability and quite specific).

At one end of the spectrum (lightly localized and highly portable) would be very generic games dealing with topics of broad interest. A game designed by one of our partners comes close to this end of the spectrum. It is an *X-Files*-type paranormal scenario, and works equally well just about anywhere. At the other end of the spectrum (highly localized with difficult portability) are games that are rooted in the history, context, and geography of a particular location. An example of a game at this end of the spectrum is one that a student of mine built about a historic battle that is played on the battlefield.

There are trade-offs inherent in the design of games along this spectrum. Games that are highly portable can be quickly transported from one place to another simply by providing new location information (maps, GPS coordinations, etc.). This makes them more easily scalable and distributable. These games must minimize the amount of local information to achieve such portability. The real world is primarily a playing board. Yet these games provide a kinesthetic and spatial component for learners to whom that appeals; the real world may just be a playing board but it is an interesting one that will have some meaning to the players because of their prior experience with the locale. It also permits much of the same interaction that the participatory simulations allow—social interaction, face-to-face communication, and group problem solving.

Games that are highly specific to a particular location can incorporate a lot of detail about the real-world location. Buildings, people, smells, sounds, and even feel can become a part of the game, allowing for tight connections between the player, the game, and the real world. A tight connection between the location and the game also provides a stronger potential for *authentic games*—games in which the activity conducted is much like a real-world practice. Games about local history, geology, populations, or the environment can be designed to take advantage of these unique assets. Much of the authenticity in these cases comes from the less tangible aspects of the real world that are inherently incorporated in these games. It is what makes the players in the augmented reality game weigh factors differently than those using a purely virtual environment. The cost is portability. These games cannot be moved from place to place without a lot of customization, or perhaps reauthoring.

I argue that the sense of authenticity, brought through the interaction with the real world, has a powerful impact on learning. Shaffer (2006) has described what he calls *epistemic games*—games designed to recreate the *reproductive practices* of particular fields. Reproductive practices are what particular fields do to train new people in those professions. It may be internships for doctors, moot court for lawyers, or some form of apprenticeship for carpenters. Each profession has those practices. To the extent that these practices are *reflective practices* (Schön 1983, 1987), they can be applied not only to train future professionals in those fields, but also to have other learners participate and learn about subject matter as if they are those professionals. Thus learning about anatomy as if you are a doctor is a more powerful way of learning that particular subject matter than learning it from a text. This does not mean that one must actually be trained to be a doctor, but participating in that reflective practice helps one learn the content. For example, students could learn functional anatomy from a textbook and diagrams, or even dissection of a rat, but they are more likely to gain a better understanding of functional anatomy if they were to role play as doctors studying medical information that related to that same content.

Shaffer refers to this discipline-oriented perspective taking as an "epistemic frame." There may be many types of experiences that can recreate such practices. For example, many schools used the Model UN simulation in classes to teach students about government and politics. When those experiences take on the qualities of games, they become "epistemic games." These are games that necessarily place the players in roles in which they must take on the reflective practices of particular fields. Through these epistemic games novices learn about the content in a deeper, more effective way. The applied fields, represented by most of the roles in these augmented reality games, fundamentally differ from the traditional "think like a scientist" activities of the classroom (Squire and Jan 2005). Correspondingly they offer uniquely interesting problems for the students to solve, and frames to view them through.

Epistemic frames can only have an impact on learning if the person sees themselves in the role that they are playing—this means the game effectively creates the simulated real-world experience for the player. If the game is unable to do this effectively, then it loses its impact. The examples earlier in the chapter showed that players in the augmented reality world felt *as if* they were environmental scientists or detectives. While it may be argued that the duration of this experience was too short to truly create an epistemic frame for this audience, it demonstrates that it came close to doing so. Consequently, through that lens, or epistemic frame, the players made decisions differently than do the players in the virtual world who were not looking through that lens. Thinking as an environmental scientist, a player might weigh the scientific, social, and political ramifications of such a decision in one way; thinking simply as a person playing a game in a classroom they might weigh factors another way.

The closer the game experience is to reality, to really feeling as if the player is a member of the designated community of practice, the easier it is to create that epistemic frame and enable learning. I argue that technologies that create experiences that vary along the light to heavy augmentation axis and the heavy to light localization axis should have corresponding variations in how authentic they are and how well they create epistemic frames (figure 8.5). A lightly augmented, highly localized game is the closest approximation to the real world and is likely to create the epistemic frames with the highest magnification. Adding more digital representation and removing the role of reality make the experience more abstract and virtual and lessen its impact.

These scales are of course relative. The augmentation required to represent any particular domain is in and of itself quite variable. Some fields may require a lot of virtual information to reasonably approximate the practices of that field. A fast-paced game tracking a source of radiation might require more virtual data than one about geology. Similarly, certain domains may require more localization than others. A game about local

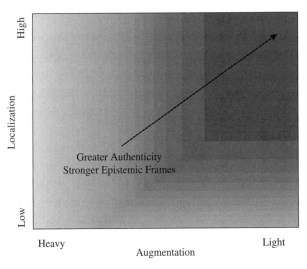

**Figure 8.5**
The two-dimensional space defined by the degrees of augmentation and localization interact to produce epistemic frames with corresponding strength.

history might need more than a game about air quality. With enough work and technology those digital representations can be quite realistic, maybe even approaching real-world realism. But the real world still has the virtual world beat in terms of feeling authentic to users, through multiple senses and intangible and serendipitous additions. As Klemmer, Hartmann, and Takayama (2006) stated, "Designing interactions that are the real world instead of ones that simulate or replicate it hedges against simulacra that have neglected an important practice."

However, there is one additional factor that arose in the nature center version of the game ED. It is not only the degree to which the game is technically localized and connected to a particular place, but also how much it connects the user to that place. While the game itself can help create that connection (it is possible to do much more than we did in this example), some of it comes only from previous experience and knowledge of that place. So the extent to which a highly localized game impacts the authenticity of the experience for the player is mediated by the player's knowledge of the place. The y-axis in the previous diagram may more accurately be labeled "Player's Connection with the Locality," which is a function of the localization.

This raises interesting design questions. Is it more effective to have a portable and less localized version situated in a place that players know well and can bring a lot of experience to? Or is it better to have a highly localized game about a place that the players have little connection to? This choice is the dilemma we faced in one of our projects on bringing augmented reality games to middle schools. We had two designs to decide between. One design was to create highly portable (less localized) games that could easily be brought to each school site. Since most of the schools in the study were urban schools with small campuses, there wasn't a lot about their locations that could be customized. Similarly, since these games were being run by teachers we couldn't ask them to put in a lot of work customizing each game to their site; alternatively, we could bring the students to a fixed site with a highly localized game developed for it. The trade-off here was that this created a field trip model, which based on past experience was less likely to have players connect with the place.

While the learning gains associated with both designs will take some years to assess, their constraints are having a strong impact on what kind of game is developed and the content of that game. Highly portable games cannot have content that is deeply connected to the place. These games are most suitable to "far" role play. These are roles that are not much like (i.e., far away from) real-life roles. The game designed by my colleagues (Chris Dede and Matt Dunleavy at Harvard University) in this case was known as Alien Contact. It was an *X-Files*-type government conspiracy game with a supernatural dimension in which students needed to use math and careful critique of documents to determine whether aliens had landed at their school. This model is similar to that used for Savannah (Facer et al. 2004), which could be dropped down on any "football pitch" (i.e. soccer field) available. The counterpart to this game, developed by a different set of colleagues (Kurt Squire and his group at the University of Wisconsin), was known as The Beach. It was set at an environmental station in an urban area to which classes could go on a field trip. Like Environmental Detectives, it involved investigating environmental contaminants with possible human impacts in a sensitive area. This game is much closer to being authentic in terms of the roles played and connections with real-life practices. Portable games lean towards being more fictionalized than do highly localized games.

The trade-off is that in the field trip model, despite being deeply connected to the game, the players are not deeply connected to the place. As a counterpoint, in the highly portable games, the games are not very well connected to the place, but the players are. Perhaps the result is no net difference, or perhaps one of these factors is more important than the other. This is an important guiding question for the future of research in augmented reality and learning.

The team enters the medical center and the doctor asks the receptionist sitting behind the desk if she has seen anyone complaining of avian flu in the last few minutes. The receptionist, more bothered than concerned, brushes off the question and continues to work. But the team is not deterred. The technician on the team responds that there is nothing to worry about, and that they are only passing through to pick up some additional masks and viral testing kits. The team hoards as much as it can find, but some personnel on the team are already overloaded with antiviral medications and other protective gear. Team members also don't find as many of the field test kits as they had hoped so they'll have to collect samples and later send them off to the lab instead.

They hurry out of the medical center and back toward the dorms where they had identified a student who had either a bad case of the common flu or perhaps avian flu. On their way there, the technician exclaims that she just noticed she is starting to feel feverish and has the chills. The group panics. Should they stay away from her? Or isolate her in quarantine? Do they need to get her medication immediately or is it best to identify the source of her illness? And what about the other patients that are waiting? Has the technician already spread the disease to others, maybe even other team members?

This scenario of a potential outbreak, or one like it, has repeatedly come to the attention of the public through news reports and even popular culture. As scary as it is, it is also intriguing and filled with opportunities for problem solving, understanding of science and medicine, and a lot of data collection and analysis. It is the kind of scenario that keeps public health officials up at night, but in the case above it fortunately is only a game. This group succeeded in many ways in the game (members had already identified someone as a possible source for the disease) and failed in others (at least one person in the group had themselves become sick). So, did this group "win" or "lose"? How do we define its success at this game?

## Quantifiable Outcomes

In chapter 3, I circumvented concisely defining what a "game" is in favor of a more functional definition in terms of what we can appropriate for classroom use. However, in designing games there are particular properties that we may seek to incorporate because they facilitate a game-like experience. For these purposes we do not want to circumvent those definitions, but rather extract the important components that allow us to design learning environments that harness these critical elements of games.

Many definitions of game include specification of "quantifiable outcomes." For example, Salen and Zimmerman (2003) define a game as: "a system in which players engage in an artificial conflict, defined by rules, that results in a *quantifiable outcome.*"

This is a fairly concise and widely held definition of a game. It is consistent with many definitions held by scholars, and perhaps more importantly it is consistent with how many students and game players define a game when asked. Note that the conflict of which they speak need not be a conflict between players, but a conflict for a player to resolve to get to some goal. The rules may also be as complex or simple as one wishes. I remember many an afternoon spent reading the rules of the Avalon Hill board games before we ever understood enough to begin playing. Other games have quite simple sets of rules, and correspondingly simple play (like Tic-Tac-Toe) while the previously mentioned Nomic makes a game out of the rules themselves. The last point, quantifiable outcomes, is somewhat openly defined. Salen and Zimmerman discuss the meaning of this criterion in the space of role-playing games (RPG), which bear a lot of similarity with the augmented reality games discussed in the previous chapters. They conclude that it all depends on the context. An RPG may not be a game in the sense that there is a single overarching goal towards which players work, but it does consist of tasks that may or may not be achieved. The success of completing or not completing each of these tasks may indeed be quantified (you either did or did not retrieve the sword from the ogre). They go on to say that RPGs may become more game-like through the addition of a competitive scoring structure, or less game-like by focusing more intensely on narrative.

Similarly, Juul (2003) includes quantifiable outcomes among the six necessary features of games, which include:

1. Rules
2. *Variable, quantifiable outcome*
3. *Value assigned to possible outcomes*
4. Player effort
5. *Player attached to outcome*
6. Negotiable consequences

So there are indeed three (in italics) of the six elements of a game that focus on outcome here, with one specifically being quantifiable outcomes. Juul further specifies that those outcomes must have some value assigned to them. This value may be explicitly stated in the rules or definition of the game, or implicitly through game play. Some outcomes are positive and some outcomes are negative. The fifth piece of Juul's definition associated with outcome is that the players are attached to that outcome. This is, as he defines it, a psychological connection through which someone might feel good or bad (or better or worse) about different outcomes. In discussing these definitions with gamers, many disagree on this latter point, that the players must be attached to the outcome. This is not a definition of a "game" they argue, but rather of a "good game." Many gamers say that they have played plenty of bad games and that they couldn't care less about whether they won or lost them. Yet by all accounts, they are still games.

I'll additionally take to task the idea that there must be values assigned to the outcomes. Or at least I'll take to task the notion that the game must do this in some way. As Salen and Zimmerman allude to in The Sims, while it is not overtly a game in that it has no explicit goals, many players turn it into a game by defining their own. What The Sims provides is a way for the players to assess their own goals. It does this through a variety of quantifiable feedbacks, an array of meters that lets you know how you are doing in a variety of categories. You have access to information on your income, savings, job, friends, and family. All of these can be used to help the player assess their success in obtaining their goal, thus making The Sims a game. This capability to use information from within the game to define goals and assess outcomes is perhaps a more generic way of defining the role of outcomes in games.

Too often, designers instead interpret quantifiable outcomes as *objectively* quantifiable outcomes, and in turn use the most straightforward definition of quantifiable outcomes as a way to incorporate this feature into their games. This means including a *score* and ways of accumulating *points*. For many genres of games, points and a score become the sole objective of the game, and in these contexts score is a useful and productive construct. In other contexts, however, score is much less meaningful or even counterproductive. In our efforts to create authentic learning environments out of these game spaces, adding a score would be at odds with the authenticity of the space. Authentic tasks require authentic assessment. This is particularly germane to the activities that we conduct, in that *defining the appropriate goals* is a key component of the games themselves. In Environmental Detectives, players needed to decide whether to protect the university's image or to protect the university from legal liabilities. On even more local scales, players needed to decide what kind of information they valued and how they thought they could maximize that information.

To learn constructively, it is essential that players define their own goals and assign their own values to those goals. In these previous games, the degree to which players met their goals could only be partially assessed by the players themselves. They knew how much information they had gotten and had some way to assess how much information was out there in total. They could also define their goals in terms of the criteria that they chose to satisfy (legal, ethical, medical, etc.). However, the missing piece was how well the plan they put in place met those goals. So a player suggesting that the university plant trees to remediate against the problem, get rid of the chemical, and protect the university's reputation had no *in-game* way to assess how successful that plan was in accomplishing those goals. Instead, that came through situating the game in other assessment contexts. In some cases this was done in role, with players presenting their case to a jury or a scientific board that assessed the feasibility and viability of their plans. In other cases this was done out of role as more traditional class presentations assessed by the teacher. These outcomes were quantified, and ideally this could be done in a predictable, consistent fashion (perhaps through rubrics). But in many ways, the quantifiable outcomes aspect of the game was weakened by the variability of this approach, and the fact that it was difficult for players to determine in-game how well they were progressing toward meeting those goals. It would be more effective if the players had feedback along the way that helped them assess or even redefine their goals, in a way similar to The Sims.

## Outbreak @ the Institute

One way to provide useful dynamic feedback to the players in the game is to couple the simple underlying models used in the participatory simulations with the authentic real-world game play of the augmented reality games. The primary challenge in coupling these two approaches is creating a consistent, coherent world in which multiple players can exist in real-time. This meant not only using underlying models to define the world players would inhabit, but also defining that world in a centralized way. Thus we defined a client-server architecture for playing augmented reality games. This created a single coherent world that would keep players synchronized in real-time, instead of the multiple parallel worlds that were part of the outdoor-based AR games.

In merging the features of the participatory simulations and the augmented reality games, we also chose to merge the contents of our past experiences with these games. We created a game known as Outbreak @ the Institute that combined the virus modeling of the Virus game, along with the public health and role-playing aspects of the AR games.

Players of Outbreak @ the Institute participated in a fictional scenario: the outbreak of an emerging disease on a university campus. Specifically, they were confronted with an outbreak of a new form of avian influenza, or bird flu, which was very dangerous because (in this fictional scenario) it had become transmissible between humans. Several students had come to campus from around the world for an international robotics competition, and some were already exhibiting flu-like symptoms. Players could encounter both bird flu and the common seasonal flu, which have very different outcomes but can be difficult to distinguish in the early stages. The players had to work together as a team in the roles of diverse professionals to gather information and use the tools available to them to contain or stop the outbreak as best they could.

This particular scenario involving bird flu was chosen for several reasons. First, it was an issue of great public interest and concern. The public is bombarded by images and articles on the spread of bird flu between birds and people, as the media plays on people's fear of a global pandemic. Interpreting this information can be a challenging task. This activity builds on that interest and hopefully provides valuable skills in making more sense of current and future dilemmas. Additionally, the community of epidemiologists and public health experts, recognizing the importance of the general citizenry being better informed about emerging diseases, provided valuable input and feedback in the design of the game with respect to models, content, and practices.

As mentioned previously, Outbreak @ the Institute differs in several fundamental ways from the previous outdoor GPS-based AR implementations. The positioning is done via Wi-Fi, in a very coarse way, resolving the player's location down to the room that they are in. Other implementations of Wi-Fi positioning (e.g., the Wi-Fi system made by Ekahau) promise much finer-grained resolution. But in practice, without a lot of specialized equipment, at this point in time there is too much variability in Wi-Fi signal strength to consistently provide better results. Like our previous games, our goal was to use commercial off the shelf equipment, so requiring additional proprietary locative technologies, or even limiting to a particular kind of network was not considered. Resolving location to the room level, however, was fairly straightforward. Each Wi-Fi access point has a unique numerical identifier. We associated each room with a "signature" of access points; that is, we identified a specific subset of access points that could be seen within a particular room. If the player's machine saw "most" of those access points then the game put them in that room. This method is highly portable and can ultimately be used by people in a variety of locations. It is much less sensitive to changes in Wi-Fi arrangements and particularly to the presence of radio frequency interference (namely people).

The wireless positioning permits much of the same interaction as the previous outdoor AR games. As players move from building to building, different virtual characters appear on the screen of their PDAs. Players can then take actions such as interviewing the virtual characters to get textual clues, video, audio, or documents. But the important difference is that the server tracks the status of all players, allowing them to interact in a single virtual world, so that the actions of one player can affect all the others. At one level, this keeps the players in a single coherent world. If a player sees an item in the room (figure 9.1a) and that player picks that item up, it will

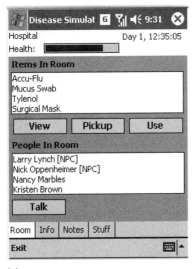

(a)

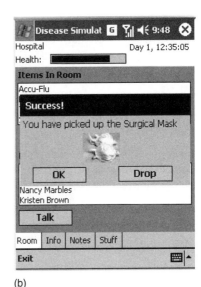

(b)

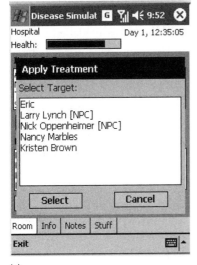

(c)

**Figure 9.1a–c**

Screen shots of Outbreak @ the Institute, showing the dynamic display of items (a) that can be picked and used by players (b). The display also shows the status of the players and NPCs (c).

not only disappear from their screen, but will also disappear from everyone else's screens. Once that item is in the player's possession they may use it in a variety of ways. In the example, in figure 9.1b the player has picked up a surgical mask and can put it on, or put it on one of the real players in the room or even on one of the nonplayer characters (NPCs). In addition to giving the players a consistent and more authentic view of the world, Outbreak @ the Institute challenges them to use a common set of limited resources. Players who choose to hoard items might be able to protect themselves, but would not help in an effort to distribute resources around so that they can be used where needed.

The constraint on resource distribution connects to another learning goal associated with limiting resources: enhancing collaboration. When items were limited only at the level of player, there was little incentive to collaborate around those resources. Constraining resources across players encourages conversation about them: Who should have the resources, where should they be positioned and who currently has them?

The most important difference that the single worldview permits is an associated underlying model for the whole world. This model allows the players to not only design solutions to the problem, but to test them as well. In this case, an underlying probabilistic model of disease transmission is used to create realistic patterns of infection. Each player and virtual character has an antigen count, representing the number of virus particles (or titer) in their body. This quantity is not directly visible to players in the game; but, as the quantity increases, the player's health level (which is visible) drops. This is one form of data that players can use to assess their progress—their own health, the health of their team, and the health of the group as a whole. Seasonal flu and bird flu have different equations governing how their antigen counts change over time. Players can infect other players, players can infect virtual characters, virtual characters can infect players, and virtual characters can infect each other as they move around. Each of these interactions is modeled by a probability dependent on their antigen counts and the amount of time they spend together in a room.

A player's status includes her current health, shown by a meter which decreases if she becomes infected with a virtual disease, and her inventory of items picked up during the game. Those items include a range of preventive, curative and diagnostic tools, as well as a variety of evidence that players might use to contain the outbreak. The player's health is affected

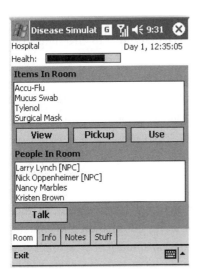

**Figure 9.2**
Game items can be picked up by players and used on other players, NPCs, or themselves.

by her disease state (modeling which, if any, disease she has and how far it has progressed), time, and medicines. Game items, such as diagnostic testing kits and vaccines, are scattered around the virtual landscape, as well as being in the initial possession of some players. These items provide specific functionality and may be restricted in use by role (see figure 9.2).

Players join the game in one of three possible roles, each of which have different abilities in the game. We chose to have three roles, since with small numbers of students in each class, we could create enough diversity in roles to encourage jigsawing of complementary information, while also providing some redundancy where information may be missed. The three roles that we chose were as follows:

Medical doctors can use the various types of medicine in the game to treat players and virtual characters. The medicines include palliatives (which reduce symptoms only), vaccines (which prevent infections), and cures (which stop the course of disease).

Field technicians can diagnose diseases. First, they use a sampling device to take a blood or mucus sample from another player or virtual character. They put the resulting sample into an analyzer, which reports the presence or absence of a disease. False-negative readings are possible due to a threshold for antigen count before it can be detected and due to error built into the analyzer, representing its inaccuracy.

Public health officials can quarantine virtual characters. There is a special location in the game representing a quarantine room, in which diseases are not transmissible. The public health officials can use a special item that allows them to transport a virtual character to the quarantine room.

### Game Goals

Players are not given specific criteria for "winning" the game. Instead, they have only the loosely defined tasks of gathering information and containing or stopping the outbreak, and a limited total amount of time to play. As a result, players must decide for themselves what their goals should be throughout the game. Allowing the students to define goal states, rather than specifying them, creates a more realistic situation in which they must evaluate trade-offs and decide on a satisfactory balance. In the group meeting at the end of the game, students are given the opportunity to discus the information gathered (for example, scientific knowledge about bird flu, and hypotheses about which virtual characters were initially infected) and estimate their degree of success in containing the outbreak based on their observations of which real and virtual characters got sick. The lack of specific criteria for winning makes the game more like real-world problem solving tasks, which are often ill-defined. At the same time students are given ample data to help them assess their own success. For example they can assess their own health, their team's health, the total number of players sick, the total number of virtual characters who got sick, the extent of the illness across individuals, the time taken to track down the initial infected person, or even how they utilized resources like quarantine tickets or medicines.

### Audiences and Authenticity

Outbreak @ the Institute has been played with a number of diverse audiences, ranging from high school students from an urban school to university students to professionals in training in a graduate school of public health. Each of these audiences entered the activity with an obvious difference in skills and background, but also with differences in expectations as to both what and how they should learn through this game.

The students from the graduate school of public health were the first to go through the activity. The students were all in a class on epidemiology that contained practicing physicians, public health officials, and nurses. They acted as a check on the authenticity of the game.

The computational models underlying the game were relatively simple, but quite accurate. We adapted current models to explain how the virus would develop within people (and NPCs) and spread between people. Each medicine and vaccine that was available also was based on currently accepted models of how they interacted with the virus load within the individuals, and subsequently how those affected the person's health and their probability of transmitting to other players. Protective equipment was also modeled to provide the level of protection that is typically associated with each of the items, which in general is less than perfect protection.

What was not explicitly modeled was the process one should use to track down and contain the outbreak. Players were given little instruction on how they should play the game, other than the technical aspects of what each player was capable of doing. They needed to figure out whether it was best to work alone, in small teams, or in one big group. They needed to figure out whether they should concentrate on minimizing the total number of people infected or limiting the extent of infection within the most advanced patients. And they needed to figure out how to protect themselves as well as their patients. In other words, we created an environment in which one learned about managing an epidemic based strongly on constructivist principles of learning.

This form of constructivist learning was somewhat foreign to the public health students. They were initially uncomfortable with having to define their own goals and their own structure for solving this problem. They stated that they would have preferred a designated leader to be put in charge who would give everyone orders and tell them what to do. That well-defined command and control center is something that they had come to expect in a crisis situation. Not having that structure put them out of their comfort zone. I would argue that this was successful in two regards. First, taking the players out of their comfort zone provided them with a good learning challenge to establish effective structures in the absence of them being provided. Secondly, it provided the players with an opportunity to construct a network for managing this situation that may be better than the usual command and control structure. Perhaps such a structure was a legacy system, a remnant of previous technologies and problems. Or perhaps the old structure was indeed better, but this provided a way to discover why it was better and appreciate the ways in which it was better.

In the end these students confirmed the authenticity of managing this crisis. When asked about the initially undefined structure for managing the crisis one of the students responded that it was "realistic. You have to 'think' about everything yourself. Life is not as clearly structured as a homework sheet with all data available." And more generally another student added, "This was first hand, like being on the ground actually investigating a hot spot (which some of us have done). More often, we learn from other people's experiences; often making the same mistakes again."

The role-playing aspect of this game took on a unique meaning with this particular group since the roles that they were playing were very familiar. Sometimes the players (by chance) played roles that they actually played in real life, while other times they played a different role. Walking a mile in someone else's shoes often brought new insights. Someone playing a doctor who was not a doctor in real life said, "I am not an MD [the role I played], so I learned how to interpret information from interviews and decide what to do. Also I learned to make sure that I was protected (surgical mask, etc.) when interviewing potentially infected people! We were also spreading the disease potentially more than it would have been otherwise." However, someone who was a medical doctor in real life and played another role felt "restrained to not have full [normal] function or ability," yet came to appreciate "the team approach since we had an MD, field technician, and public health official in a group."

Many of the students commented on one particular aspect of the realism and authenticity of the game, namely that it was much harder to apply what they had learned in textbooks and lectures when they really felt like they were dealing with a crisis. "I liked the feeling of 'real life'—getting an idea about the difficulties of logistics, limited resources, importance of communication and coordination." In fact some of the students found that the situation sometimes felt "chaotic" (or like "herding cats" as one of the students described) and that many of the other players were making poor decisions that would in turn affect them. "It was interesting to see how under some stress people just wanted to start quarantining anyone with symptoms! Points to the importance of protocols!" This is exactly the kind of learning that the game was designed to create. Rather than specifying protocols and having the group take them for granted, they began to learn that such protocols were important.

The chaos that many of the players talked about was tightly linked to the players' own concerns about getting themselves or their teammates sick. The problem changed from something abstract that they could handle using their knowledge, to something personal that they cared deeply about. So, with some verification of the authenticity we could examine how students use this environment to set and assess their own success. We could see how students balanced goals, and how those played out over time. In other words, we could see whether these students could use the feedback from the models to make this a game.

## Assessing Success

As might be predicted, many students entering the game set fairly clear goals of "containing the disease" or "making sure no one gets sick" or even "learning about bird flu." We asked students to define their goals at the beginning of the game, when they had been briefed about the scenario but not yet played the game at all, and then again at the end of the game. They were given another opportunity to define their goals and then rank them in order. At the beginning of the game their top three goals were:

1. Learn about bird flu.
2. Learn about strategies for responding to a disease outbreak.
3. Keep myself from getting sick.

These first two goals may have been influenced by what students expect from an activity in school: to learn about content. Those goals, however, would be hard to assess in the context of a game. After playing the game, their goals had shifted from content-based learning goals to a more practical and easier to assess set of goals:

1. Keep myself from getting sick.
2. Keep my team from getting sick.
3. Find out who was sick first.

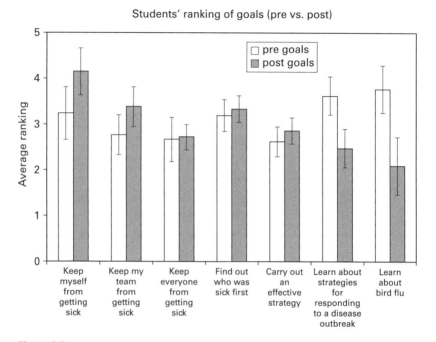

**Figure 9.3**
Students' ranking of goals before and after the game show an increased preference for self-preservation.

These revised goals were easy to assess—their own health and the health of their immediate teammates. They learned that these are practical considerations: if you don't have your health what do you have? They also focused on one particular piece of data that they determined was practical —who was sick first, referred to as Patient Zero by epidemiologists. Identifying Patient Zero is one of the most critical pieces in helping to contain an outbreak. This was not taught to them explicitly in the game, but working through the network of contacts, and finding out how quickly it had spiraled out of control led them to conclude that they should really have that piece of information.

When asked why they shifted their goals, one of the students responded: "I think, just going based on memory, I could be wrong but I think I was more so focusing on strategy [initially], and keeping people from getting sick. But more so today I was focusing on keeping myself from getting sick, keeping my team from getting sick. I think keeping my team from getting sick was a lower priority than it was this time."

The student added, "I think it was because when we were all together, it was kind of like, you're getting sick, oh no, I have to help you, and then also, like, wait, you're getting me sick, too! So, you know that part of it became an important thing. And then also as we were interacting with more and more people, if we found people who weren't sick, you know that was fine, but then we found people who may be sick, like, they were going to be around like a big crowd of people, then it became more important to keep them from getting sick, because then it would spread even faster."

This student has shifted from a noble goal of keeping *everyone* from getting sick to a more personal goal of keeping *themselves* and *their team* from getting sick. This is partially out of wanting to protect themselves, but also comes from an understanding that if they are sick and moving around a lot and interacting with other people, then they could be spreading the disease far and wide. This is reinforced by another student who added: "I think it was more like I saw myself getting sick, well me and [my friend] Krystal, and we cared more about ourselves. Then when we tried to get [sic] her from spreading...we just kept getting sicker and sicker. So treating ourselves was more important."

So while these motives may be looked at as selfish (protecting yourself) they also may be seen as critical in helping to solve the problem. Protecting yourself is also one expression of an understanding of the connectedness of the system, and the importance of single individuals. This isn't just about a system, it is about a collection of people, any one of which could be highly important. As one student said, "I didn't really know how like one person going to one place with so many people could affect so much. Like I knew it but I didn't think it was like really [important]." This is a fundamental understanding of the dynamics of the system, and an indication that students are using the feedback from the system model to help them redefine and assess their goals.

So did this ultimately help the students define their own success? And what was that definition? We asked the students if they were successful, and they were able to point to specific indicators of that success. Said one: "Yeah, I think we were fairly successful, because the one person, because, we only found one person who was sick. But when we found them, we told the doctors, and the doctors got them medicine, and we also got out of that room, so we wouldn't get sick. And then, um, and we ourselves, when we got sick, we told the doctors to you know give us vaccine, but we kept getting even sicker, and we didn't understand why, but once we left the room, and they gave us a vaccine again, we were fine. So they treated us again. So I guess maybe because we were in the room with sick people it still wasn't helping, I don't know. That could have just been in my head, I don't know. But, so, we were pretty successful because we only dropped maybe halfway in our health, then we got healthy again. And then you know there wasn't like a large outbreak of people, we only found one sick person and then we got that under control, so I think we were pretty successful."

The student here has defined success in terms of measurable numbers of people who got sick, their response to them, and the impact on their own health. They have also indicated that they have begun to understand the dynamics of the system ("because we were in the room with sick people").

At the same time, many of the students were cognizant that there were limits to their success, that they could have done better in the game. Remarked one student: "[Were we successful?] I mean, yes and no. We never did find the start of everything. We only found one person that was infected. There could be more people that were infected. So, that's all I know."

One unexpected theme in this group's assessments of success were measures of how well they coordinated, collaborated, and communicated with each other. As one student put it: "Yeah [we were successful], it was like we were working with the other groups, and then, we're just like able to communicate and work together to find out what's happening and what to do about the NPCs."

This idea of communicating and working together was pervasive throughout many of the player's definitions of their success within the game. That definition connected strongly with the idea that the players were successful if they played the roles well. They were successful when they were working like doctors, technicians, and public health officials. One student said: "I think we were mostly successful because right after we figured out how to analyze the blood tests and everything, we would do it to everyone and ourselves. And then since we were with the doctors we told them [the players in other roles] right away to give us some vaccines, and then we got better. So I think we were pretty successful...because we worked together. So we would give them the masks and the doctors would vaccinate them or give them a pill or something."

So their success is measured by how well their performance matches their expectations of the role itself. The game play is motivating them to achieve a more effective epistemic frame. They are striving to better meet the expectations and practices of the role that they are playing.

While the game doesn't directly report how well the players are fulfilling their roles, it does provide evidence that they can use on how they respond to situations, how they use their resources, their own health, the health of their team, and the health of the NPCs—these factors are all dependent on how the players are responding to their roles and to the situation.

These responses can also be observed in the behavior and language of the players in the midst of handling the crisis:

Rebecca:  I'm going to start interviewing people since I'm the doctor and that is what they do.

Miguel:  Harry is a SARS researcher and he has a low fever and is sweaty.

Rebecca:  Does it say what he researches?

Steve:  Did you notice that there are masks here? Respirators.

Miguel:  Should we use them on ourselves or other people?

Rebecca:  Let's use them on Harry.

Steve:  He may or may not have the illness that we are concerned about.

Rebecca:  Why don't I interview them since I get the health status [when I receive interviews].

Miguel:  But you might get different things [in your interviews], since we're different roles.

[Incoming over walkie-talkies]:  Iago just tested positive: we [public health officers] quarantined him.

Rebecca:  Does someone have a test kit? We should test Harry.

[Incoming]:  Iago should be quarantined and Cindy should as well.

Miguel:  Can we call for a field technician to do a test on Harry?

This exchange shows that the students are appropriating the aspects of the roles that help guide them in their behaviors. They understand the interrelationship and interdependency of their roles. The doctor knows what she needs to do in the situation (interview patients), and the public health officials (PHOs) radio in that they too have taken the actions that are associated with their role (quarantine). Finally, they don't have a field technician present but realize they need one to analyze samples. These actions are what team members will later use to assess whether or not they have done their job successfully.

**Quantifiable Outcomes Revisited**

Looking back to Juul's definition relating to outcomes, it includes *"variable, quantifiable outcome," "value assigned to possible outcomes,"* and *"player attached to outcomes."* The model underlying Outbreak @ the Institute allowed the players to define their own quantifiable outcomes, typically in terms of their own health, the health of their team, and the health of NPCs. Combining those metrics with their use of resources, and the response to that use (in terms of health and spread of disease) gives them good evidence to quantify their success within the game. There is clearly some value preassigned to the different outcomes—getting sick is bad and staying healthy is good. But through the game the players ascertain what being "completely" successful might be. Is it possible not to get sick at all? Is getting sick, catching it in time, and getting better just as good? The players clearly take on the ability to assign that value. Finally, the players are indeed attached to the outcomes. From the players' rankings of their goals, they have shifted from a series of meta-level goals to concrete and very personal goals. They don't want to get sick, and they don't want their friends to be sick. This may be only a game to the players, but that is saying a lot.

There are few clues left behind. Whoever did this knew what they were doing, and what the police would be looking for. Still, everyone leaves behind a trace. And there are only so many buyers for stolen artwork, even a classic such as this one.

Police officers are everywhere, hitting the streets to take statements from witnesses who may have seen something—a strange car, some unusual noises—anything out of the ordinary. Other officers seek experts who can tell them about potential buyers of this artwork, as well as how the thief or thieves might have committed the crime. Forensics experts are combing the scene for fingerprints, and scanning for fibers from clothes or even an errant hair that may have been shed by one of the perpetrators. Those samples are sent back to the lab where experts in microscopy examine the fibers and attempt to do DNA analysis of biological samples that they have found. Computer analysts are scanning databases trying to match details from this crime with other recent crimes, criminal records from recent criminals, and the origins of the one piece of equipment that was left behind. All of this information is collected in a central repository that is accessed via phone calls, secure Internet access, and mobile devices in real-time. The first twenty-four hours are the most important, and with six of those gone by already, this massive team needs to work quickly.

In another corner of the globe, a group has assembled just outside a cave. They too have a puzzle to solve. In this case they're trying to figure out if they're going to retrieve the sacred Orb from the depths of the cave that is known to be inhabited by a herd of orcs (nasty monsters).

One of the scouts has returned, and unfortunately has not had success in finding alternative entrances to the north. He did, however, see evidence in the form of fresh footprints that at least some of the orcs may be out for a hunt. A while later another scout returns; although she has not found an alternate entrance, she *has* found what appears to be a weak spot in the cave that could be penetrated by someone with digging experience. Two dwarves head back out with the scout to investigate that possibility, taking along with them several others with battle experience in case they are successful. In the meantime the rest of the group prepares for a frontal assault. Two guides using cloaks will lead the way, one down the path to the east and the other to the north. The goal is to try to pick off the orcs one by one, for the group doesn't have a chance if the orcs counter-attack en masse. Each orc will be stunned by a magician of the group, immediately pelted with arrows by an archer, followed up by hand-to-hand combat. If anyone finds evidence of the Orb they will communicate their position and the group will join them.

While both of these narratives are fictional, they represent the kinds of collaboration and role specialization that are required in real-world practices (the first narrative) as well as in a virtual world of role-playing games (the second). These players could have been engaged in tracking down the source of an emerging disease, locating a problem on a power grid, or optimizing the design of a new car. It isn't just large tasks that require such skills. In fact, players could have also been determining a slowdown on a company's intranet, prepping a pitch for a new ad campaign, or planning a new office building. Many of the challenges and problems that we face at work and in our lives must be solved collaboratively. While the Information Age has allowed each person to become increasingly specialized in what they do, at the same time they must be able to adapt and interact with a diverse range of people with other skills.

So how does the raid on the orc cave compare with modern challenges in the Information Age? In massively multiplayer online role-playing games (MMORPGs), people must adopt these same techniques. Each player develops particular specialties that makes them unique and valuable on a team (Jakobsson and Taylor 2003; Steinkuehler 2006). These games, such as World of Warcraft, Everquest, and Lineage, require players to form teams (or guilds) that are assemblies of players with the right mix of skills to solve the problems of these worlds. Each problem (e.g., slaying a monster, protecting a castle, retrieving artifacts) is much too difficult and immense to be tackled alone. Instead they require players to form these guilds, many of which are long-lasting, and build relationships and team skills. The game does not specify how the problem must be solved or what skills might be required. Instead, the guilds themselves must determine the right skill set and how to coordinate its members (which often number in the dozens).

Given the importance of collaboration in the real world, as well as the benefits of collaborative learning, experts are starting to explore the learning that goes on in MMORPGs. There is evidence of team building, social network formation, and expert communication (Steinkuehler 2004; Johnson 2005; Yee 2006) that emerges in these worlds, not only during the intense critical moments of such games, but also in the spaces in between when players are working to form their teams and their relationships. Other experts are starting to explore how MMORPGs can specifically be applied in education, creating MMORPGs for learning (Dede et al. 2004; Eustace et al. 2004; Barab, Warren, and Ingram-Goble 2006; Galarneau and Zibit 2006). For example, Barab's Quest Atlantis provides a multi-user virtual environment (MUVE) in which middle school students take on socially minded quests while learning academic content.

In a related example of a MORPG (multiplayer, though not "massive"), my colleagues created a game called Revolution (figure 10.1), about Colonial Williamsburg on the eve of the American Revolution. This game was an extensively "modded" (i.e., customized through a toolkit) version of Neverwinter Nights, a classic medieval role-playing game. The goal of Revolution is to either get the revolution to occur (if you are on that side) or to stop it from occurring (if you are on the other side). Each of nine roles in the game (including blacksmiths, carpenters, and even slaves) must decide what their interests are (based on their role descriptions) and how they can contribute to the cause. The main weapon that each player possesses is information. They talk to each other and to NPCs to convince them of their case. The game is only "complete" when one player of each role is engaged, as they each hold a critical piece of the puzzle.

## Collaborative Learning in Augmented Reality

The goal of many of our AR games has been to capture authentic learning opportunities from engaging in games in real spaces. Given the importance

**Figure 10.1**
The MORPG, Revolution, required players to collaborate as they worked to influence the American Revolution.

of collaboration in real-world practices, our priority is making sure that the game is structuring collaboration is critical to its success in achieving these goals. Additionally, the game may be both more enjoyable and a more positive learning experience if it is structured around collaboration, as in the style of MORPGs.

Barab, Warren, and Ingram-Goble (2006) describe the educational MMORPG Quest Atlantis as blurring the boundary between what is game and what is real (Salen and Zimmerman's [2003] "magic circle"), by bringing real-life content into the game world, and considering the impact of game events on the real world. This form of play keeps players in their "Zone of Proximal Development" (Vygotsky 1978), using the playful nature of the world to structure tasks that are just beyond the edge of the player's expertise.

In our AR games we strive for a similar goal, blurring the "magic circle" even further through play in the real world. As seen in some of the previous examples, players incorporate much of their real-world surroundings into their game play. This structure allows learners to tackle difficult real-world problems and tasks in a playful manner. Sitting between MMORPGs and real-world collaborative problem solving, AR games occupy a perfect

**Table 10.1**
Collaborative Learning Components

| | |
|---|---|
| Positive interdependence | Group members perceive that they are linked with each other so that one cannot succeed unless everyone succeeds. |
| Promotive interaction | Students promote each other's success by helping, assisting, supporting, encouraging, and praising each other's efforts to learn. |
| Individual accountability | Each individual student's performance is assessed and the results are given back to the group and the individual. |
| Interpersonal and small-group skills | Students develop the interpersonal and small-group skills required for an individual to function as part of a team. |
| Group processing | Group members discuss how well they are achieving their goals and maintaining effective working relationships. |

place to incorporate collaboration as a core game dynamic, further blurring the magic circle between game and real-world practice and learning. However, our initial attempts at fostering collaboration in AR games were not particularly successful. Over successive iterations of AR games, we performed design research experiments to enhance the in-game collaboration through a stronger emphasis on role playing and more sophisticated game dynamics.

The evolution of the role of roles within our AR games demonstrates the pivotal nature of this component of game design for learning. In examining this evolution, we adopt a collaborative learning framework (Johnson, Johnson, and Holubec 1994), which specifies positive interdependence, promotive interaction, individual accountability, interpersonal and small-group skills, and group processing (see table 10.1), as the necessary components of collaborative learning.

It is of note that many MMORPGs match well with this set of criteria. The basis of such worlds is the ability to communicate and collaborate with other players to solve problems. These worlds simultaneously rely on collaboration and build collaborative skills. Ducheneaut and Moore (2006) describe the situated learning inherent in MMORPGs such as EverQuest Online Adventures. There is a strong sense of communities of practice (Lave and Wenger 1991) being built as new players enter the game and work along the periphery while more experienced players teach by example and coaching. Specifically, they look at four opportunities for situated learning in MMORPGs:

- In-game, in-context discussions:   communication within the game within roles

- Out-game, out-of-context discussions:   communication between players outside of the game world

- Observation:   players watching each other within the game world

- In-situ teaching:   experienced players explicitly teaching novices

In order for groups within these worlds to be successful they must take advantage of these learning opportunities to train incoming players. Table 10.2 shows how the learning opportunities in typical MMORPGs align with the criteria promoting collaborative learning.

Our initial AR game (Environmental Detectives), while promoting some collaboration, didn't draw on enough of these skills and consequently didn't match up particularly well with these requirements. As a result of

**Table 10.2**
Collaborative Learning in MMORPGs

| | |
|---|---|
| Positive interdependence | Missions in most MMORPGs are not goals that can be accomplished alone, but must be done in concert with several or many other players. |
| Promotive interaction | Players need to boost each other's skills so that they are successful in accomplishing their individual tasks. Players' fates are tied together and they must work with together to promote their respective skills. |
| Individual accountability | While missions are collective, rewards are solitary. Each person individually collects treasures and experience points. |
| Interpersonal and small-group skills | Players must be able to communicate and listen, as their characters' lives depend on it. While communication may be spoken or written (often in shorthand), it is important that each player understands the others. |
| Group processing | Hordes are transient, breaking up and reforming with new players as necessary. In regrouping, they must decide who to keep and who to replace based on past experiences. |

refinement and redesigns, successive generations fare better through an increased emphasis on role playing and game mechanics, borrowing heavily from the elements of successful games.

## Generation 1: Environmental Detectives

The problem space of Environmental Detectives (ED) is quite vast. By design, no one player can obtain all of the requisite information in the allotted time, and teams had to work with one another to collect data and come up with solutions. Each team consists of two or three students supplied with one Pocket PC, one walkie-talkie, a printed map, and a notepad. Teams typically assigned one player to the Pocket PC/map, and another one or two players as notetaker and/or communicator. This promoted strong collaboration within teams—forcing players to work together effectively for navigation and planning. In most cases, players were not specifically instructed to either collaborate or compete with the other teams in the game, but to use their judgment in order to devise the best solution and provide the strongest evidence. By creating this large physical space, which can easily be geographically subdivided, we were most strongly emphasizing positive interdependence. It should be noted that in Environmental Detectives, there is no "role" differentiation among players and, since all players are using the same software, they can potentially access the same information at the same time. Table 10.3 depicts the ED components designed to promote collaboration.

Many classes have run through Environmental Detectives in many locations (see chapter 7). Here I come back to some of that data to examine collaboration within and across teams in several classes, in particular looking at one class that had previous experience and training in collaborative problem solving and one that did not. Through these runs we have found that collaboration within the teams was quite strong. Both groups of students collected interviews and used sampling, though how much weight they gave to these qualitative (interviews) and quantitative (sampling) activities varied greatly.

### Experience without Prior Collaborative Problem-Solving Training
Reexamining one team of three students toward the end of their investigation, we see how they start to evaluate what they know, and they

Table 10.3
Collaborative Learning in Environmental Detectives

| Promotive interaction | Moving in physical space, students working collaboratively can cover more ground and share information by looking at each other's screens. One group's information is often evaluated on the spot by other groups. |
| --- | --- |
| Individual accountability | Each pair of students is responsible for presenting their case to the class at the end of the experience. This is often supplemented by written arguments. |
| Interpersonal and small-group skills | Groups of students communicate via walkie-talkie to share information or pool data. |
| Group processing | A classroom/lab space provides a shared location where students can plan their next steps, assemble evidence, and ultimately present their case to the class. |

grow concerned that they do not have enough information to make a compelling case about the toxin. We pick up the discussion as they decide what to do next:

Louis:   My socks are so wet.

Camera:   We should head back soon.

Gina:   Yeah, it is 12:50.

Louis:   How far away is the thing [place they should return to]?

Gina:   Where do we have to go again?

Stacey:   Alan Morgan center? That is . . . .

Louis:   [Looking around.] Not around here.

Stacey:   Right here [points at paper map].

Stacey:   How are we supposed to make recommendations?

Gina:   I don't know.

Louis:   Just read off of the information that we got.

Gina:   I thought we could dilly-dally but we actually did work.

Louis:   For once.

This group is representative of the somewhat superficial experience that students had when treating the investigation as a scavenger hunt common to many field trips. Perhaps influenced by the field trip nature of the experience, some students thought the goal was to acquire as much information as possible and then develop the right answer. Here, late in the experience, they begin to understand that developing an answer requires negotiating and synthesizing information. Although this group shared a lot of information and fluidly navigated multiple information spaces, much of their collaboration centered on game mechanics, and less on collaborating to work through scientific dilemmas.

### Experience with Prior Collaborative Problem-Solving Training

Another class with previous experience in collaborative problem solving divided the problem space and worked together to efficiently solve the problem. The facilitator began by asking students what they knew and what they needed to know, and asked them to make a plan. The groups went out and collected data, and, mid-game, decided to pool resources and see what they had learned so far.

At the beginning, the game unfolded much as the others had, although there was considerably more sharing between groups. In order to encourage more reflection in situ and strategizing in situ, we planned time for the teams to regroup and share information halfway through the activity. As the groups came back in the room, the teacher asked the class who took samples and who took interviews. Joan (a student) responded by asking if they could collaborate as one big group.

Joan:   Did everyone write an "x" where they dug their holes on the map? Do you think we could put the whole on the map to show where we dug them?

Stacey:   Do we have an extra map?

Sue:   We could use one person's map and they could become the designated map. Or you could redraw it on the giant post-it.

The group started passing around a paper map and marking down where they dug wells.

Sue:   Maybe there are some people who might want to come up with a master plan now. What are we going to do with this data? Do we know everything we need to know? And then how are we going to prioritize it before we go out there? Do we have a report yet? Do we have a sense of a report? Do we even know what's going on with the wells?

Joan:   I have a map with readings on it. I noticed that ours were a lot higher than everyone else's . . . and in one of our interviews it says that as the chemical disperses the readings get smaller. So we could probably see where it spilled.

Sue:   Do you have a guess based on—

Stacey:   I have the map . . . and it looks like down where—

Joan:   Our highest reading was like 81 and we have 50 something and everyone else got below 20 mostly.

Stacey:   Also, I was just reading everybody's notes, and it says they used it in building 3, and building 3 was in our section, so I'm thinking maybe while they were using it they could have spilled it there. But, that's just a guess.

Sue:   Okay.

Joan:   Maybe they, um, tried to get rid of it through the pipes or something and it went into the ground there.

Having raised several questions, two girls stood at the board and added to the list of facts already started earlier in class. A map was passed around the room, and students noted where they had found their toxins. To those familiar with knowledge-building communities (Scardamalia and Bereiter 1991) or jigsawing (Brown and Campione 1996), the scene was quite familiar. Each student was adding what she knew toward building a more holistic view of the problem. What was particularly noteworthy about this session was how the facilitator, Sue, quickly receded into the background. While she drove much of the initial conversation, the students shortly assumed responsibility for organizing classroom talk.

The group continued sharing information for about ten minutes. A student, Miranda, interrupted the conversation:

Miranda:   Before we go back out, can we go through the names of all the interviews and make sure that everyone's hit one at least?

Alice:   From our interview I didn't...we didn't get the name of the person, but I think [the virtual character] said his name was Harold.

A few girls walked over to Alice's table and showed her how to review her interviews. Joan called out from across the room how to review the interviews they had already seen. As they reviewed, Stacey re-read the "About TCE" resource and started reading aloud. Sue, the facilitator, instructed the class to listen, as this might be important. Again, the girls assumed responsibility for driving activity, and the facilitator's role was to help the class attend to its own productive insights.

Eventually, a handful of girls returned to the field to collect more data. They were particularly interested in repeating measurements for increased reliability. Using walkie-talkies, groups in the field communicated back and forth with the classroom, which functioned as a control center. "Take a reading by building three," commanded one girl over the walkie-talkie to the others in the field, who seemed to enjoy taking suggestions on where to drill for wells in order to collectively beat the clock. Although the class had started out divided into groups, by this point it was essentially functioning as one big group with smaller subteams self-organizing around subsets of the main problem.

When they returned to the classroom, the class again aggregated information. The facilitator was nearly silent, as the girls polled one another to see who was tracking the flow of TCE, if anyone found out where TCE was used, and so on. The class began to piece together a causal explanation:

Monica:   We can pretty much tell that the spill is in this area [pointing to map].

Joan:   Also, the numbers are higher on the bottom because one of the interviews said that it slopes down to the river

Stacey:   So maybe it's up here and been there for a while and that's why it's sloping down.

The facilitator reminded the girls that they needed to structure their report for the university president. The girls walked through their solutions, using charts of paper to illustrate their ideas. In an important move, the facilitator asked the girls to present their findings as pairs (as they were assigned), but the class declined. "We'd rather just present as a group, if that's okay," Joan said. So, for the next ten minutes, the girls collaboratively made their presentation, with each girl contributing different ideas and facts.

The differences between these two cases, the one with collaborative problem-solving experience and the one without, shows that Environmental Detectives is structured to allow collaborative problem solving, but not necessarily facilitate it. It is not obvious to teams that have not conducted such activities before how they can structure their collaboration. They all *can* get the same information, so it isn't obvious why they must work with each other. The game itself doesn't offer a lot of hints as to how they should collaborate or why they would need to. However, it does have a lot of opportunities for collaboration, as demonstrated by the second group, if players know what to look for. It is a wide-open, unstructured space that can be partitioned geographically and conceptually. There is ample opportunity to collaborate, if one knows what to look for, but the game could be structured to scaffold that collaboration more significantly.

## Generation 2: Charles River City

While the collaboration *within* groups was strong and successful in the first-generation AR games, the collaboration *between* groups was limited or nonexistent, except in the last case, which showed that promoting collaboration at a larger scale requires providing additional scaffolding for collaborative learning. In order to promote greater collaboration between the teams, the core engine for our AR games was redesigned. From this redesigned engine new games have been created. Primary among these is Charles River City (CRC), which combines environmental science and epidemiology to create a large-scale investigation.

Charles River City, loosely based on Chris Dede's MUVE River City (Dede et al. 2004), is MIT's second-generation outdoor GPS-based augmented reality game. While the game itself introduces some key new game-play and learning elements, it is also built on a dynamic engine that will allow others to build similar games using an authoring tool that should be released in the coming months.

The basic scenario for Charles River City is that there has been an outbreak of illness coinciding with a major event in the Boston metro area. The event changes for each run of the game, so that it is based on something timely. One of the first runs started out like this: "The July 1, 2004 headline of the Boston Globe reads '26 More Fall to Mysterious Illness as DNC Looms.' A rash of disease has swept through Boston; and—with the Democratic National Convention coming to the city in a few weeks—citizens, politicians, and health officials are all concerned. What is the source of the illness? Is this an act of bio-terrorism or a naturally occurring event?"

Players are then told that a team of experts is being brought in to investigate the problem, including epidemiologists, physicians, public health experts, and environmental specialists. This group must work together first to evaluate case reports and available surveillance data, then to plan and implement rapidly an investigation to determine the cause and source of the outbreak, assess risk, communicate with the professional and public communities, and identify effective interventions. The team will compile information through collection and analysis of environmental samples, hospital records, patient histories, clinical samples, and testimony from community members, then engage in collaborative analysis and interpretation. The team must determine its findings and propose actions very quickly in order to assess the risk, propose timely means to reduce risk and treat affected persons, diminish societal fears, and work with decision makers to design and implement a solution to the problem.

While much of the game play is similar to Environmental Detectives (and in fact all of the capabilities of ED are also built into the newer engine), including location sensing, interviewing, and data collection, the new game offers several features that make it a more interesting and dynamic environment. Notable enhancements designed to foster collaboration include:

Distinct player roles   Players take on one of several distinct roles in the game. These roles provide the players with special capabilities (e.g., being able to take certain kinds of samples or receive unique information). The player's role also dictates what information they get from the NPCs (e.g., someone might say one thing to a police officer and something entirely different to the medical doctor).

Increased role of data beaming   Players can beam data that they collect to other players within the game. This information might be interviews that they have collected from disparate places or differing roles, or it may be field data that they have collected.

Cascading events   Events can cause the triggering of other events. For example, speaking to one NPC might cause another NPC to appear somewhere else on the map (i.e., an NPC tells the player about someone else to interview). The triggering event may be something that a player experiences directly or may come from information beamed to them by another player.

Distinct player roles added several key elements. First, players receive different information from virtual characters depending on what role they are playing. For example, a virtual character who is feeling sick might give a player in the role of nurse different information than she would give to a player in the role of detective. Second, roles have different data collection capabilities allowing players to collect unique types of samples or access unique kinds of data according to their role. For example, an environmental scientist might have access to water sampling equipment, whereas a medical doctor might be able to access medical records or get vital signs from virtual characters. Finally, since different roles can access different

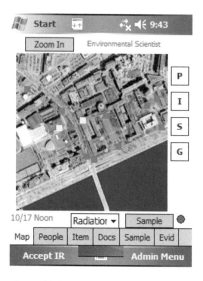

**Figure 10.2**
A game screen from Charles River City shows the player's location on a map relative to NPCs and samples.

information, players can use infrared beaming to exchange information between players. For example, in CRC, a character reveals information to the detective about a student who has fallen ill. The detective must then share that information (via infrared beaming) with the nurse, so that the nurse can interview the player and examine the specific symptoms and what might be causing them. Reconsidering the criteria for promoting collaborative learning, we see how these new game-play elements have enhanced the potential for larger-scale collaborative learning.

In subsequent iterations of AR games, we have found these new features to be effective at fostering collaboration, which in turn scaffolds a more authentic investigation process. The fact that sharing information could reveal new things encouraged frequent digital exchanges, which were accompanied by pertinent discussions of game progress. Here is a typical exchange of middle school students from an urban school in the Boston metropolitan area playing CRC. In this particular version, there are three roles—a doctor who can take people's vital signs and symptoms, an environmental scientist (Env) who can take samples from the water and air, and a department of public health official (DPH) who has access to hospital records and epidemiological data.

**Table 10.4**
Collaborative Learning in Charles River City

| | |
|---|---|
| Positive interdependence | Each team's information is explicitly described to them as only a small piece of the puzzle, and they need information from other roles to solve the problem. This sharing is facilitated by the infrared beaming of information. |
| Promotive interaction | Students encourage players in the other roles to go out and get information that they know they need but cannot get themselves. |
| Individual accountability | Each role has access to unique information necessary to solve the problem. Players know which role has access to the information that they need. |
| Interpersonal and small-group skills | Sharing of information across teams via the infrared beaming becomes a point of instruction on how to share information and collaborate across teams. |
| Group processing | Groups "divide and conquer"—with players in their specific roles dispersing to find information and regrouping to exchange information with the other role players at planned times or ad hoc, or to move around in multirole groups. Both strategies require getting together and planning how to move forward. |

Manny:   I got a document that says that says West Nile Virus has the most serious effects on people over 50.

Jane:   So...the doctor might be the one that wants to talk to Salvadore [previously identified older patient] since he can get his health information.

The doctor went to the location where Salvadore is and took a physical exam. A player in the DPH role also went along.

Sal [Doctor]:   I was right! He has all of the symptoms of WNV [West Nile Virus].

Tricia [DPH]:   [Radios to whole group via walkie-talkie] I found Salvadore!

Sal:   [Via walkie-talkie] He has all the symptoms that he carries for WNV.

This collaboration between groups continued on into the classroom where they made their recommendations on what to do about the problem. Each one contributed information that they got on the topic specific to their role.

Dave [DPH]:   I found that West Nile Virus can make you really sick.

Tricia [Env]:   Mosquitoes are all over the world so it is dangerous.

Manny [Env]:   Not right now. Since it is fall there aren't many mosquitoes out. Only in spring and summer.

Jose [DPH]:   Julia Washington [a virtual character] said that an elderly man complained of swarms of mosquitoes.

Kim [Doctor]:   We found the old man [Salvadore] that complained of symptoms that could be WNV.

Dave [DPH]:   There might be enough mosquitoes where it could still be a problem.

Manny [Env]:   So get rid of the ones that are there.

Dave [DPH]:   Get rid of all of the water that is standing around like in old tires.

Manny [Env]:   And tell people to wear long clothes when they are outside.

As seen in the above dialogue, the different roles have different perspectives and different pieces of the puzzle. This encourages them to collaborate, which progresses into other forms of collaboration and discussion as they attempt to solve the problem at hand.

### Generation 3: Mystery @ the Museum

The use of handhelds for learning is not confined to the more formal environments of K–12 and higher education. Museums have sought to employ handhelds to engage visitors and learners more deeply and broadly across their exhibits. Some museums have built upon the ubiquitous audio guides found at many museums, and have started offering handheld devices that allow visitors to specify exhibits for which they would like to subsequently access additional media. Perhaps the best known example of this is the Experience Music Project (http://www.emplive.com/visit/about_emp/tech.asp) in Seattle, which not only provides supplementary audio content to exhibits, but also allows users to electronically "tag" items that they can then explore in more detail using a separate electronic workstation at a later time. Other museums have also sought to offer electronic guides to visitors that not only provide supplementary information on the spot, but also allow them to retrieve related information later. The Exploratorium in San Francisco has conducted a study (Hsi 2003) of visitors' use of this strategy within their science center. In this study, location-aware Pocket PCs provided visitors with Web-based information about aspects of the museum including history, annotations, and suggested explorations. Content, including audio, video, and text, was delivered to the devices wirelessly. Two themes emerged in this study. First, visitors said that the technology isolated them. In order to hear audio they wore headsets, which tended to separate them from their surroundings. Additionally, visitors tended to focus on the device, taking away their focus from the rest of the museum. Second, visitors had trouble connecting the virtual content on their hand-helds with the real content in the museum. Despite these two shortcomings, however, the visitors did say the technology encouraged them to view exhibits in new ways and try things that they hadn't before.

Museums have not employed these new technologies to encourage interaction with other museum-goers. For the most part, as noted in the Exploratorium study (Hsi 2003), the technologies do exactly the opposite, fostering a more private and isolated experience. Yet the field of computer-supported collaborative learning certainly provides evidence that collaborative learning is effective in encouraging people to think critically about important ideas, and perhaps this notion should be more seriously considered in the informal learning space of museums.

Building on our experiences with the Environmental Detectives game at informal learning spaces like nature centers (Klopfer and Squire 2003), a new game was designed for the Boston Museum of Science. In choosing a target audience for the game, we settled on the core museum-going contingent of families—specifically late elementary through middle school-aged students and their parents. A primary goal was to increase meaningful collaboration and interaction between parents and children around science and inquiry. These goals are consistent with the recently introduced American Association for the Advancement of Science's (AAAS) supported Science Everywhere initiative (⟨http://www.tryscience.org/parents/parent .html⟩).

In parallel to the previously mentioned efforts using AR simulations outdoors, we have developed another platform to create similar experiences indoors. This platform replaces the GPS-based positioning outdoors with Wi-Fi positioning, allowing less precise but adequate "room level" positioning indoors. At the same time it allows the players to remain in a wirelessly networked environment, which comes with the additional benefits of being able to synchronize actions with a server and communicate across distances, though those capabilities were not used until more recent games.

The fictitious premise of Mystery @ the Museum (M@M) was that a band of thieves (The Pink Flamingo Thieves) left their calling card (a pink flamingo) in an exhibit case indicating that they had stolen a priceless object from the museum and replaced it with a replica. The players of M@M were brought in as a team of experts to try to solve the crime, apprehend the criminals, and identify and retrieve the stolen artifact. Each player took on one of three possible roles—a technologist, a biologist, and a detective—each with special capabilities. The interdependencies among the roles encouraged players to collaborate throughout the game. Logistically, players were organized as six players (three pairs) per team with each pair (parent and child) using one Pocket PC and a walkie-talkie.

The Pocket PC used Wi-Fi positioning to determine in which room of the museum it was. It could then provide the players with information about dynamic virtual characters and objects in the room with which they could interact. These virtual objects and characters in turn referred to and complemented real, physical components of museum exhibits, which had been incorporated into the story. The fundamental interactions that were inherent to the game were as follows:

• In each room was a set of *virtual characters*, which could be "interviewed" by clicking on them. The characters would provide a monologue in the form of text, often accompanied by pictures. The characters could move rooms over time, and players in different roles might receive different information from the characters (i.e., a character might tell something quite different to a detective researching a case than to a biologist). Many of the virtual characters referred to other exhibits or rooms.

• In many rooms there were *virtual objects*, which could be picked up and examined. Each had both a textual description and one or more images associated with it. Players could also "show" virtual objects to characters who would then react accordingly, often providing additional information. Some of the objects related to nearby exhibits.

• In several locations *virtual equipment* (e.g., a SEM, a scanning electron microscope) could be used to obtain further information about the virtual objects. Where possible the virtual equipment was placed near real equipment of similar types (like the SEM). Equipment "use" was restricted to certain player roles as appropriate.

• Several items in the museum were tagged with *infrared tags*. These tags provided the players with virtual samples taken from those particular items (e.g., fingerprints from a glass case).

- Players could exchange objects and interviews with each other through *localized infrared beaming* (like on a TV remote control). In many cases one role was the only one capable of retrieving a sample (e.g., the detective who could retrieve a splinter from an unconscious guard), while a player in another role could use equipment to analyze it (e.g., the technologist capable of using the virtual SEM).

The game was completed when players had accumulated enough evidence to obtain a virtual warrant for the arrest of the culprits. One of the organizers played the role of judge who considered the information presented orally by the players and, if sufficient evidence was presented, beamed the players an arrest warrant.

Several of the elements of this game enhanced the potential for collaborative learning through increases in key areas. The most significant is the increased role of unique skills and information given to the players, which play a key role in this game. Additionally, the use of the space itself can play a significant role in increasing collaborative learning, giving real artifacts a role in the game. Players must interact around these real artifacts, combining powers across roles, and often running into other teams in areas laden with information. Together these changes result in a greater potential for collaboration.

M@M was played at the Boston Museum of Science on two successive weekend afternoons with a group of approximately twenty parents and children each day. Parents were always paired with their own child. While several of the parents and children knew each other, the majority did not know any of the other participants before the game. The groups were subdivided into teams of six (as mentioned above). In cases with uneven numbers, a single redundant role was added to a team. After players were introduced to the "mystery" and given a brief tutorial of game mechanics, they were given one hour to play the first phase of the game. After this first hour of game play, players regrouped in the meeting room, checked in with the organizers for five to ten minutes and then went back into the exhibit halls to play the second half of the game for an additional thirty minutes.

The roles in the game turned out to be extremely effective in engaging the pairs of participants with one another. Each individual role was forced to collect and share information to successfully solve the case. Here one group has met up after collecting information separately:

**Table 10.5**
Collaborative Learning in Mystery @ the Museum

| | |
|---|---|
| Positive interdependence | Roles receive unique information, and can perform unique tasks. Many tasks explicitly involve stages during which players of different roles must participate and interact. |
| Promotive interaction | Teams are assigned to work towards a common goal. Frequent interactions around a succession of clues involving each role builds confidence and interaction. |
| Individual accountability | Many clues require the action of unique players on the team, requiring each member to take care of their jobs. |
| Interpersonal and small-group skills | Sharing of information and items across teams via infrared beaming is critical to solving the mystery. Many clues must be collected by one team member and processed by another, require negotiations around data collection tactics. |
| Group processing | Coordinating the presence of each of the team members around critical pieces of information requires planning and logistics. Coordination can take place using walkie-talkies that blanket the entire game space. |

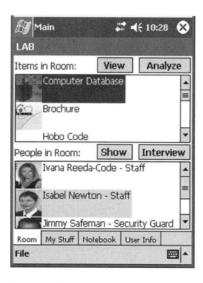

**Figure 10.3**
A game screen from Mystery @ the Museum shows the NPCs and items that are "in" the room with the player.

Boy 3:   Have you been to the mummy?

Mom 2:   Yes we went there.

Boy 3:   They have to go there since they're the biologist. It is upstairs—

Boy 1:   Let's give you [the technologist] the splinters so you can look at them with the microscope.

Mom 1:   We got the hobo code but we can't fully decode it. What do you think this means? [Beams to other groups so that they can all look at the picture.]

Often the groups concluded that it was beneficial to move around the museum in groups that included multiple roles so that they could collaborate to solve the problems. As one parent said, "In the second part we all went together to every room. Even though we might not have needed everyone in each room we did better as a group." One of the senior museum educators further commented, "sometimes people have trouble with the logical reasoning . . . [but in this group] they saw that one person could get what the others couldn't and they got the power of roles. Then they started using the beaming and they got that roles idea and off they went." The interdependence of roles served as the starting place for collaboration, which then promoted more general collaborative problem solving. It is interesting to note that in the post-game surveys many participants wrote that they felt that their role was the most important in the game for one reason or another. This was consistent across all of the roles, showing that they had fostered players' sense of a unique contribution in addition to promoting collaboration.

Players in the game were required to visit a wide variety of places in the museum, and to examine exhibits closely to find and understand some of the "clues." Several codes, for example, were woven into the storyline (the thieves used codes to communicate with each other). Interpreting these codes required players to find and connect information from several exhibits on mathematics, communication, and models. The feedback from the participants suggested that this combination of depth (examining some exhibits in detail) and breadth (thinking more broadly about multiple exhibits) was engaging and effective in encouraging them to think about the museum's exhibits. This can be seen in the interactions of one of the groups searching for information to help them decode one of the clues that the thieves left behind:

Mom 1:   We're looking for codes to help us decode this. If anyone finds stuff let us know [looking around].

Girl 2:   Over here! Over here!

Mom 1:   [Boy 2] Look in the 14th century [points to chronological history of mathematics].

Boy 2:   Look Look. Water and dice like on the code.

Dad 2:   [Reads information about the code to himself and then applies that to the code "written" on the back of a virtual receipt.] In an … hour … [points to a part of the exhibit and speaks to the group] … it is telling him when to meet by the water. An hour after close.

This interaction shows how the teams worked together to discover and apply information from real exhibits in order to interpret the virtual information, which in turn fed back into their game-play strategies. During the group debrief discussion following the game, these feelings of connection between the real and virtual contexts were further conveyed.

**Generation 4: Outbreak @ the Institute**

In the previous chapter I presented a detailed study of Outbreak @ the Institute, the first client-server-based game that we created. That chapter specifically investigated the role of underlying dynamic models and feedback on game play. But another element of that game was increased reliance on roles and collaboration. Like the other AR games, players in Outbreak are equipped with handheld Pocket PCs as their link between the real world and the virtual world of the game. The Pocket PCs receive location information based on simple Wi-Fi positioning. Ubiquitous network coverage enables the Pocket PCs in the game to stay connected to a server. Unlike the other games mentioned, this allows all of the players in the game to live in one common world where the actions of one player have immediate (and delayed) effects on all of the other players. For example, a player might pick up a virtual item in a room, which would then not be accessible to any of the other players. Similarly, a player might contaminate a scene, move important items, or interact with an NPC in ways that have consequences for other players. This fundamental change in the game engine has opened up new opportunities of game play and learning exploration. Rather than assess the situation and recommend solutions, as we have done in Environmental Detectives and Charles River City outdoors, players must actually contain the problem. Among the actions that they can take are:

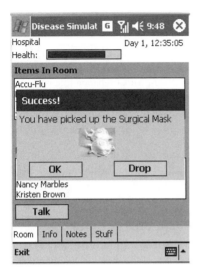

**Figure 10.4**
A game screen from Outbreak @ the Institute shows a player picking up a Surgical Mask. That item will consequently disappear from the screens of other players in the room.

• Take and analyze samples from real and virtual players in the game to test for the presence of diseases.

• Obtain and provide medicines to real and virtual players in the game to control the spread of the disease.

• Use preventative gear (e.g., masks and gloves) to control the spread of the disease.

• Quarantine individuals to control with whom characters come in contact.

The game engine shares some common features with previous games, like the ability to take interviews and samples, but also adds virtual limited objects that have specific purposes. Perhaps most importantly, it connects *all* of the players in the game through a common server. This means that through both intentional and unintentional actions the players affect each other and affect the overall outcome of the game. This stimulates whole-group collaboration out of necessity, and fosters thinking about collaboration more generally. This is further facilitated by in-game tools for messaging in which players can send messages to each other through a bulletin board and instant messaging tools, allowing targeted collaboration

**Table 10.6**
Collaborative Learning in Outbreak @ the Institute

| | |
|---|---|
| Positive interdependence | Medical diagnosis and treatment require the participation of multiple roles. One role can diagnose the disease and a separate role is required to treat it. |
| Promotive interaction | Teams are assigned to work towards a common goal, and the game inherently binds everyone to that goal. Players coordinate across the entire group to try to reach that goal. |
| Individual accountability | Most often the groups divide and conquer, with each subteam responsible for treatment in particular locations. Within those teams, each member has further responsibility for carrying out their assigned tasks. |
| Interpersonal and small-group skills | Sharing of information through walkie-talkie, instant messaging, and bulletin boards requires the building of a number of communication skills. Players need to effectively use those tools, and communicate within their subgroups to accomplish the goals. |
| Group processing | Groups are coordinated through the communication tools previously mentioned, and must share their success with other groups throughout the process. At the end, players are required to assess their own success. |

between two individuals and massive coordination among whole groups. Finally, the whole process of testing patients and giving them the medicines required was designed to require multiple roles, necessitating the interactions of players for some of the basic game operations.

The strong dependence on collaborative problem solving was apparent to players. When asked about the indicators of their success in the game, they often pointed to collaboration as the key. One student explained: "Cause in the end, when we're all just working together, it was like we're all doing this and that to help each other out and then just trying to find out what's happening and everyone would just share information."

Another student added, "I didn't realize how important it was for us to talk to each other until we actually did this." This demonstrates that these games not only build collaborative problem solving skills, but also foster reflection on these ideas, a necessary component of transfer to new situations.

## The Next Generations

One of the key design considerations for the different roles within the AR games is how much overlap there should be between the roles. Too much overlap will remove the positive interdependence and individual accountability that encourage collaboration. However, too little overlap does not give the students enough common ground to discuss the problem space. We have found that when students access the same information, it serves as a promotive interaction—reinforcing students' sense that they have done well. It also gives them a point on which to begin discussion. They start piecing the puzzle together around the common pieces and then work towards their own unique contributions. As in the CRC example (generation 2), all of the students learned that West Nile Virus was a serious mosquito-borne disease, but only certain roles were privy to the seasonality of the virus, its current levels, or the symptoms of an individual who might actually have the disease. There is still much to be learned through investigating this overlap among roles to determine how increasing and decreasing overlap can affect learning outcomes, and how role interdependence relates to both subject matter and student experience.

# 11 Learning to Write without a Stylus

Writing is one of the most heavily emphasized practices in American schools today. As evidenced by the focus of the No Child Left Behind Act, learning to write is one of the most important skills to be taught in school. But defining what "learning to write" means is another story.

In many cases, learning to write means being taught the rules of grammar and syntax, the proper form for an essay, and many other technical details of writing. While these skills are undoubtedly necessary for effective communication, they are certainly not sufficient. These skills alone do not enable *complex communication*, defined as "conveying not just information but . . . persuading, explaining, and in other ways conveying a particular interpretation of information" (Levy and Murnane 2004). Communication in the twenty-first century demands much more than mastery of the basic rules of grammar and syntax, perhaps more so than ever before in history.

There is wide recognition of the value of these communication skills, and English and Language Arts standards will often incorporate elements of these skills at their core. For example, looking at the "Guiding Principles" of the English and Language Arts standards from the state of Massachusetts (Massachusetts Department of Education 2001), we see that one of the ten guiding principles is: "An effective English language arts curriculum emphasizes writing as an essential way to develop, clarify, and communicate ideas in persuasive, expository, narrative, and expressive discourse. At all levels, students' writing records their imagination and exploration. As students attempt to write clearly and coherently about increasingly complex ideas, their writing serves to propel intellectual growth. Through writing, students develop their ability to think, to communicate ideas, and to create worlds unseen." (p. 4)

This principle clearly focuses on the content and message of the communication, not just the form. It points to *persuasive* discourse (among others) about *complex ideas*, though it does solely point to written work in this regard, rather than oral expression or expression with other media. However, those alternative forms of communication are highlighted in several of the other Massachusetts guiding principles:

• An effective English language arts curriculum develops students' *oral language and literacy* through appropriately challenging learning. (p. 3)
• An effective English language arts curriculum provides for literacy in *all forms of media.* (p. 4)

Further, the guiding principles point to some of the component thinking, research, and analysis skills that are required to perform complex communication:

• An effective English language arts curriculum develops *thinking and language* together through *interactive learning.* (p. 3)
• An effective English language arts curriculum teaches the strategies necessary for *acquiring academic knowledge*, achieving common academic standards, and attaining *independence in learning.* (p. 5)

Fully half of the ten guiding principles have nothing to do with the mechanics and technical details of writing, but instead pertain to developing the skills for complex communication using varied modes of research and communication. Yet like many of the other "back to basics" calls associated with NCLB, even when there is recognition that these other skills in complex communication are desirable, it is assumed that we cannot teach those skills without first teaching the basic rules. In practice this is a fallacy. There is nothing to suggest that people can't learn the skills necessary to *persuade, explain*, and *convey interpretation of information* either prior to or in parallel with learning the basic rules of grammar.

Devising ways to teach these skills in persuasive communication, or rhetoric, are not trivial, as it is not simply a matter of "stating your case." The communicator needs to understand their audience, listen to and evaluate other arguments, assess information that they gather and how it fits with the rest of their case, and give weight to complex and sometimes competing arguments from sources with varied credibility. Additionally, learning about effective rhetoric requires the learner to self-reflect on their own statements and arguments. This is indeed complex.

## Situated Language

It is important to note that we do not learn to read and write generically, but rather we learn to read and write in specific contexts. That is, our language is specific to the topic, content, audience, and place of communication. Reading a note from a friend about plans for the weekend is quite a bit different than reading a paper in a scientific journal on advances in genomics. Gee (2004) argues that schools incorrectly focus on the abstract mechanics of learning to read and write instead of learning to read and write in academic content areas. Not enabling students to read and write in these academic content areas (and in fact often alienating them from those areas) locks them out of thinking about many technical areas. Without the language to understand advanced topics, students struggle with them and disengage.

Gee further argues that students can learn these specialist languages if they are given the opportunities to do so. He cites the example of how young children develop vast vocabularies and language constructs around the game Pokémon. Young children are able to express complex ideas about this game because they become immersed in the Pokémon experience, and become experts in its language over time. More generically Gee states that: "The human mind works best when it can build and run simulations of experiences its owner has had (much like playing a video game in the mind) in order to understand new things and get ready for action in the world. . . . People learn (academic or non-academic) specialist languages and their concomitant ways of thinking best when they can tie words and structures of those languages to experiences they have had—experiences with which they can build [mental] simulations to prepare themselves for action in the domains in which the specialist language is used (e.g., biology or video games)." (p. 4)

Thus the foundational building blocks for helping novices build the skills to communicate in specialist areas and develop *expert thinking* in those domains are authentic experiences in those domains in which learners acquire and use new ways of reading, writing, and thinking. Additionally, when the learner can picture those experiences and replay them as "mental simulations" they have acquired a reusable tool with which to further think about those domains. Augmented reality simulations offer such a foundation by creating rich authentic experiences, which both provide and require the use of specialist languages along with physical simulations that are easy to recall.

## Scientific Argumentation

Thinking about and constructing rigorous arguments around scientific issues involves a unique set of challenges. When discussing issues grounded in science, one must understand, at least rudimentarily, the scientific principles involved as well as how they apply to the arguments being constructed. Discourse of this type is rare or altogether absent in most schooling (Kuhn 1999). Students do not have the training or experience in scientific discourse. Yet engaging in discussion around everyday issues of broad concern (e.g., global warming and stem cell research), requires such skills. The question is how can such discourse be integrated into the experiences of students.

## Augmented Reality, Participatory Simulations, and Rhetoric

Throughout our work with augmented reality (AR) games we have found that these experiences, which are situated in the real world, have been associated with players making nuanced decisions guided by both in-game information and context that they have discerned from their surroundings. The cases that players make for the actions that they have taken are complex and informed by a wide variety of information sources. In conveying those rationales to their audience (whether it be a mock jury or a real teacher), players have shown signs of thoughtful consideration of the recipient of their message, as well as for the role that they have played during the game. They have viewed the situation at hand as a persona that combines their own thoughts and values with those of the character they are playing. Related work (Squire and Jan 2007) has shown that the context of AR games, situated in familiar surroundings, can indeed be effective in promoting scientific argumentation for students.

Given this rich suite of dynamics and decision making, we have worked to create a new set of AR experiences that intensely focus on rhetoric within the game. To do this, we drew upon previous mobile technology-based activities (I won't usually call them games) that also are explicitly focused on rhetoric, debate, and reflection. The first of these activities, simply known as Discussion, was created by my colleague Susan Yoon (Yoon Forthcoming) and used the previous platform for our participatory simulations, the Thinking Tags, or Badges. The Badges (figure 11.1) were developed by researchers in Mitchel Resnick's Lifelong Kindergarten group at the MIT Media Lab. The initial Virus game research was also conducted using the Badge platform (Colella 2000). The Badges were very simple computers that could display a single two-digit number on their screens, as well as red, orange, or green lights on any of the five LEDs on the front of the unit. Input by the user was done through either a dial on the side of the Badge or one of the two buttons on the back. Most importantly, users could send and receive directional infrared from unit to unit at close range.

While the interface, display, and embedded programs were all simple by requirement, this simplicity defined the notion of participatory simulations as a technology that supported learning through peer-to-peer interactions without being heavy on the technology. Participants spent most of their time interacting with each other, as the Badge itself offered little but a way to interact with other people.

Yoon's work used the Badge to support structured discussion and debate among middle school students. Seeking to develop students' abilities to construct and understand complex arguments around correspondingly complex issues, she used the Badge to simply display what a person's current thoughts were on an issue. A person could dial in their current opinion (from −5 strongly against to +5 strongly in favor, with zero representing neutral) on the issue at hand. The LEDs on the front of the Badge would change to reflect the currently dialed-in opinion. Red LEDs represented an opinion against, and green LEDs represent opinions in favor— allowing learners to quickly scan the room to get a sense of the current feeling on the issue.

During the Discussion activity, students would walk around and interact with other students by beaming them their opinions. They could then discuss the issue, and argue why they felt the way they did about it. The Badges tracked the player's opinion and all of their partners' opinions over time, giving them an artifact around which they could reflect—thinking about what influenced them, and what influenced their partners.

In the unit that Yoon researched, the statement that the students discussed was "I believe that genetically engineered foods are beneficial." Over the course of several classes students engaged in debate about this

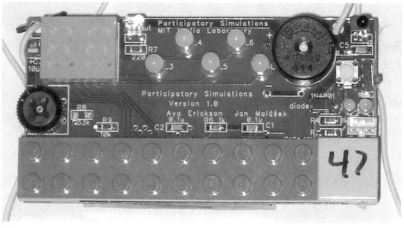

(a)

(b)

**Figure 11.1a–b**

Picture of the front (a) and back (b) of the Thinking Tags, also known as Badges, the initial platform for participatory simulations, and the platform for research on the Discussion game.

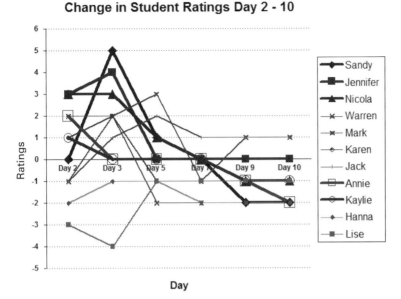

**Figure 11.2**

Changes in opinions of students in the Discussion game. Republished from Yoon (2007).

topic, noting their opinions using the Badges, and tracking their rationales on paper.

Yoon found that the students' opinions were highly variable at the beginning (and became even more extreme in the early stages), Yet as the students developed a better understanding of *all* of the issues, they developed more moderate opinions backed up by much more complex rationales (measured using scales of complex systems thinking—see Yoon for details).

While it would be a stretch to attribute the change in opinion (or the change in complexity of rationales) to the technology, this simple technology did enable the students to reflect on their own and other's opinions. Further, it provided some artifact for reflection for the students as their opinions changed over time.

The public display of information is an enticing feature of the Badges, but ultimately the limitations in terms of cost, display, and complexity favored the use of ubiquitous devices like PDAs over this proprietary platform. One research study compared the use of the Badge and PDA

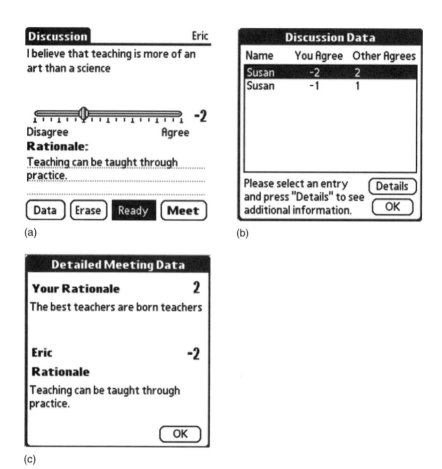

**Figure 11.3**

The PDA version of Discussion, which permits (a) entry of opinions, along with tracking of (b) history and (c) rationales.

technologies (Klopfer, Yoon, and Rivas 2004), and found little significant difference between the platforms for participatory simulations. Thus, later versions of Discussion were implemented on PDAs (figure 11.3). The transition to the PDA enabled students to more readily access the history of their interactions, and even enter rationales (though full text rationales were later exchanged for merely entering a numeric code associated with a category of rationale, as full rationale entry on the PDA was cumbersome).

When Discussion was revised and ported to the PDA, a slightly different variant was also added, one in which the answer was not merely an opinion, but a quantitative estimation. This version of Discussion was designed to be used with so-called *Fermi questions* (named after the scientist Enrico Fermi who was quite fond of these types of questions), which involve quantitative estimations and several layers of rationales. These are the kinds of questions that have become popular during interviews in the technology and consulting industries. For example, "How many tires are there in the United States?" or "How many bags of peanuts would fit in a 747?" These questions involve a combination of calculation, estimation, and thinking across several scales. In order to answer the question about peanuts on a 747 passenger plane a person might first estimate the size of a bag of peanuts, and the volume of a 747. But then they would need to take into consideration how well the bags pack, the space taken up by seats and equipment on the plane, and the ability to stick peanuts in the overhead compartments.

In a study (Klopfer and Groff 2007) implementing both forms of Discussion in classrooms, we found that the PDA version also facilitated reflection on students' opinions. It is noteworthy that we integrated some classroom dynamics that while not explicitly integrated into the technology, did support a more game-like environment, awarding points to students for the accuracy of their estimates and ability to influence others. We found that the temporal patterns of student interactions differed across the two variants of Discussion. While both versions echoed Yoon's results of student opinions converging over time, the trajectories of these changes were different. For the Fermi questions, students seemed to individually gather information for the first few minutes, and as a result their own estimates did not correlate with the estimates of the students with whom they met. In the final minutes a very different pattern was observed, where students' estimates correlated very highly with the estimates of those with whom they met and interacted. For the opinion questions, the pattern was reversed; there was a stronger correlation between student opinions' and peer interaction during the beginning of the game, with students being less influenced by those with whom they met later on.

The results of these activities suggest that integrating similar opinion tracking and reflection opportunities into AR games could provide a rich context for development of rhetoric skills, while simultaneously supporting understanding of complex issues at the intersection of science, technology, and society.

## POSIT

In order to have real-time information about the global dynamics of a class or group, an AR game centered on rhetoric and discussion needed to be built upon a networked infrastructure. Thus POSIT (Public Opinions of Science using Information Technologies) was born from the fusion of the networked AR platform used for Outbreak @ the Institute (chapter 9) and Discussion. In creating this new platform we considered which audiences might be interested in developing skills in these areas and fostering under-standing of the kinds of issues that we would be implementing. This included both formal education environments and informal education. Much of the design of POSIT was informed by discussions with museum educators at the Boston Museum of Science (particularly Barbara Costa who runs The Technology Forum) and collaborators at the MIT Museum, including Director John Durrant. One of the challenges in educating the public about emerging issues in science, technology, and society is that there are few places to convene adults to learn about and discuss these issues. Science museums and centers are perhaps the only places where this happens with any regularity and we seek to capitalize on these venues through this technology.

The first implementation of POSIT was built for the MIT campus in con-junction with the MIT Museum, and was designed to be used with multiple audiences including high school students, college students, and adults. The scenario for this version of POSIT is a fictionalized story based on a real di-lemma. At the time, a level-4 biosafety laboratory (BSL-4) was being pro-posed to be built in Boston. BSL-4 is the highest bio-safety level, required for the study of the most deadly pathogens. There are numerous reasons why the location, in proximity to universities and biotechnology research, is particularly compelling. But there are also many reasons not to put a BSL-4 laboratory in a densely populated area, making this a rich topic for discussion based in location, interests, science, and argumentation.

The fictionalized version of the scenario had a similar facility being built on the MIT campus on a site that had been cleared for another building, but was still vacant. The group gathers at the beginning of the activity and is presented with the question "Should MIT build the BSL-4 laboratory on this site?" Before the players enter their opinion, they are assigned one of ten roles in the game, and are told that they are not entering their own opinion, but that of the character whom they are playing. These characters range from university students living near the site to a nurse at an area hospital to a parent that lives in the neighborhood. For example, the assistant professor in biology is told: "You are in your third year as a professor, and just settling in. You have a spouse and two young children, and you hope to maintain a normal family life in addition to your work. You do research with infectious agents at the BSL-1 safety level."

Many of the descriptions point either to no particular interest in the lab, or a conflict of interest, as is the case with the biology professor who could benefit from the work, but seems to have other considerations at hand. Other descriptions point more simply to one side of the issue, such as the local parent who receives the following: "You live on the edge of the campus on the tenth floor of an apartment building. You have two small children, one and three years old. You like to take them out to the local green spaces to play."

Each player is provided with an electronic dossier containing this information for their character that they can read before they enter (figure 11.4) their initial opinion in the role of that character.

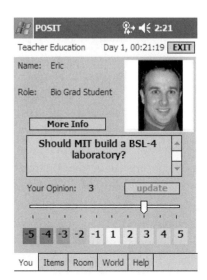

**Figure 11.4**
The opinion entry screen in POSIT, showing the player's role and current opinion.

After each player enters their opinion they begin the actual game. In the spirit of the other AR games, players walk to various buildings on campus in order to gather evidence. Their handheld device detects which building the player is in and displays the relevant game content. Each building represents the actual place, and the physical construction site on campus represents the proposed lab site. Players can communicate with each other face to face, and also through instant messaging on their devices.

A variety of virtual characters representing a range of opinions are distributed throughout the buildings in the game. They are situated in realistic locations (e.g., there are virtual students in the actual dorms and virtual nurses in the actual medical center). Players can "interview" these virtual characters to get their opinions (in text messages) on the controversy. Virtual items such as newspaper articles, journal articles, technical documents, informational pamphlets, photographs, and advertisements are distributed among these locations as well. News flashes, text messages, and other bulletins arrive at fixed times, to a subset of players according to their role. This dynamic content is used to create story lines that develop through the course of the game.

Players can select the most persuasive evidence they have gathered (items, announcements, and responses from virtual characters) (figure 11.5a) and add it to their "evidence portfolio" (figure 11.5b). The portfolio has limited storage so players must carefully choose which evidence to put into their portfolios. Once it is populated with evidence, the portfolio serves two purposes within the game (in addition to forcing the players to weight their own evidence). First, players use the evidence portfolio when trying to persuade other players of their opinions. When players meet and want to engage in dialogue they press a button that transfers their portfolio to their partner's computer. This becomes the basis for their conversation, and the players must both defend their evidence and use it to justify their opinions. They may also use it to target their opponents in the hope of changing their opponents' opinions. Secondly, the portfolio can be used to change the opinions of nonplayer characters, who will in turn change what they say when other players talk to them. As a result of these interactions the flow of information within the game can change.

For example, an NPC graduate student who players find in the library is initially inclined to be against the BSL-4 laboratory. When this NPC is interviewed initially, it gives this response: "I'm a grad student here and I

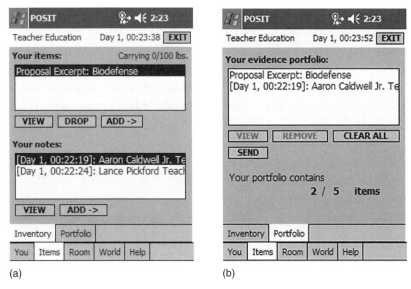

**Figure 11.5a–b**
Screen shots in POSIT showing the evidence that players can collect (left) and the portfolio that they use to organize their evidence (right).

do a lot of social science research so I spend a lot of time in this library and I know my way around the periodical section. I heard that in MIT's proposal for the new biohazard lab, they were claiming that in eighty years no BSL-4 lab had had any accidents or releases. I found that a little hard to believe so I came in here to see what I could find. Sure enough, there have been various incidents. For example, all in 2003 there were a couple of anthrax releases at Fort Detrick, Maryland; researchers in Taiwan and mainland China were infected with SARS, which spread to others; a security failure due to power outage in New York; missing samples of the Plague in Texas; not to mention a handful of personnel infections of some other diseases. Some of these were not reported to the public until many months after they happened, so just think of how many more there could be that were not disclosed. And then I hear there's info about the Soviet anthrax leak that will be revealed tomorrow over at the humanities library…when does it end? I don't know how many people MIT thinks they can fool, but I am not one of them. Whether they build this death trap or not, I'm at least going to be aware of all the risks."

This character is clearly biased against the lab and offers up several pieces of information that others could use in their case against the lab: individual cases of accidents or releases, which are very convincing (and authentic), as well as a reference to another document (about Soviet anthrax) that the players can access in a different location. There is also language that is loaded with emotion ("death trap") that may also bias a player reading the dialogue.

However, if players are successful in convincing (based on a model that follows) this NPC otherwise, he will change the tone and detail of his message: "I'm a grad student here and I do a lot of social science research so I spend a lot of time in this library and I know my way around the periodical section. I heard that in MIT's proposal for the new biohazard lab, they were claiming that in eighty years no BSL-4 lab had had any accidents or releases. I found that a little hard to believe so I came in here to see what I could find. Sure enough, there have been a few incidents in various places, I mean nothing is perfect right? But the labs in those places did a lot of important work as well. It is a little worrisome, but I still have faith that MIT and the firm designing the lab know what they're doing. If they think the benefits outweigh the risks, I'll be on board."

In order to provide consistency with information that other players have previously obtained from this NPC, the initial portion of the dialogue remains constant. But the rest of the information changes. Gone are the references to the specific cases and the upcoming availability of further information at another location. The language is much more tempered (stating that it is "a little worrisome" rather than a "death trap"). Finally, the reference to the benefits outweighing the risks gets back to a central issue that many of the players debate.

In this version of POSIT the NPCs have a rather simplistic model underlying their decision making. Like real players, NPCs start out with an opinion. In the case of NPCs that opinion has been preprogrammed based on their characteristics. NPCs also have access to their own interviews, and any documents or information that they give to other players.

Each piece of evidence in the game (every interview, document, and artifact) has an inherent *persuasiveness*, which is unseen by the players but programmed into each item. The persuasiveness is an indicator of how much that item might be predicted to influence someone's current opinion. These values are based in part on the design of each item and how much it was intended to influence players' opinions, and in part on observation of how much the items actually influenced opinions. In general the persuasiveness score is less than one, as it takes quite a bit of information in practice to sway someone a whole point.

These values could in theory be revised over time, based on statistical analysis of their influence on players in practice, but in trials to date these relationships are difficult to determine since players do not always immediately adjust their opinions after receiving new evidence. Instead, they think about it for a while and adjust their opinions when the time is right. For NPCs, however, the effect is immediate. When the NPC receives an evidence portfolio they first throw out any information that they originated or that they have already heard, assuming that only new information can be influential. A sum of the persuasiveness is then calculated. Each NPC is endowed with a *suggestibility*, alternatively thought of as resistance, to new information. So an NPC with a zero suggestibility will never change its opinion, whereas an NPC with a suggestibility of one will take evidence at face value and use the inherent persuasiveness.

Finally, the NPCs adjust their current opinion based on this new numerical input and check to see whether they have crossed a threshold from one of the preprogrammed responses to another. For example, in the graduate student text above the split was simply positive and negative. If the grad student currently had a negative opinion it would say the former response, and if its opinion turned positive it would say the latter. Splitting into positive and negative was the most typical arrangement, though one could add a midrange response as well.

## Gaming POSIT

While the opinion dynamics and responsive NPCs have the potential to make POSIT an interesting experience, it still falls short of feeling like a game. The basic elements of gaming are there in the form of motivation through the portfolio exchange and a fundamental goal of making a decision about the question at hand. Feedback about the global dynamics of opinions supports the players' quest to fulfill that goal (figure 11.6). Players have access to live updates of each player's and NPC's opinion, as well as the location of each of those players and NPCs. This information can be used to assess the player's own impact on the current opinion and also to

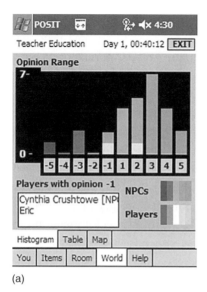

(a)

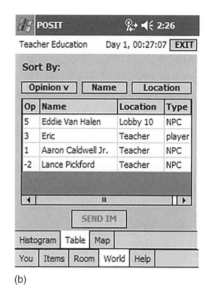

(b)

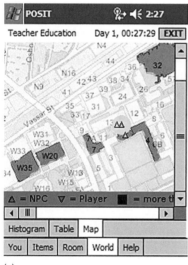

(c)

**Figure 11.6a–c**

Screen shots showing the global information accessible to players during the game of POSIT display live updates in the form of (a) a histogram of all players' and NPCs' opinions; (b) the opinion of each individual player and NPC; and (c) a map indicating the location and opinion of each player.

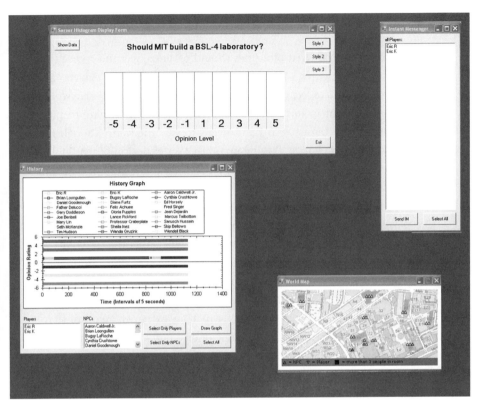

**Figure 11.7**
Live readouts along with data recording track each player's opinion to be used in both research and for the players' own reflections.

strategically target specific players or NPCs to influence or obtain information from. For research purposes this is also collected live on the server (figure 11.7) and saved for analysis by the researchers and for reflection by the participants.

This system of feedback and goals provides the basis for game play, but early tests showed that players rarely exchanged portfolios with other players and did not have a sense of the game progressing. What resulted was a system that stagnated, and players' interest in interacting diminished as time went on. In order to enhance the game play, and advance players' interest in interacting and discovering additional evidence, we added two additional supports: intriguing plot lines that developed in real-time over the course of the game, and a rating system combined with competitive team scoring.

It may seem obvious that plot lines should progress over the course of a game, but POSIT (figures 11.6 and 11.7) is designed to have players take time to consider and reflect on the information that they collect. We would not necessarily want to have players working against time trying to hit a moving target. The big issue at hand needs to stay the same, though the details within could change. Picking up on that latter notion of manipulating the details, we decided to introduce a parallel but relevant subplot that *did* advance in real-time.

One of the central issues of POSIT is the trustworthiness of information that players obtain. No one in the game is expected to be an expert on the topic, and instead they must become experts in evaluating the credibility of their information by understanding its source, verifying the information with secondary sources, and even analyzing the language of the information. As information abounds on the Internet, these skills have become highly prized twenty-first-century skills, which are extremely relevant to research and writing. Playing on this theme of credibility, the subplot centers on an incident that if verified would seriously jeopardize the safety record of the institution, but if a hoax would embarrass some of the institution's opponents.

The subplot is introduced in this manner: early in the game players receive a "News Flash" on screen warning them that "Diseased Mice are on the Loose." The alert states: "All day yesterday people reported seeing diseased mice with various symptoms running through the buildings of the campus. The earliest sighting was reported at 10 a.m. and reports became more frequent throughout the day. The infected mice are generally believed to have escaped from one of the biology labs that conduct experiments with mice, but this has not been confirmed. No effort has been made as yet to catch the vagrant mice because no one wants to get near them."

NPCs provide some information relevant to this subplot, and new characters are introduced who are mentioned by the other characters but never met "in person." Additional news flashes provide updates on the situation, and most of the evidence points to the incident being a hoax. This includes a mass email that the players receive from a professor, who states that he has evidence that this incident was a hoax and even points his finger at a potential culprit. But the true origin of the mice is never stated.

This subplot added a fun sense of narrative through the game, and early on players paid a lot of attention to the incident. In this sample dialogue from one of the evidence portfolio exchanges between two players, the mice were a central issue:

Angela: How do you know the mice getting out of the facility weren't being let out by someone? How do you know they escaped? How do you know that someone didn't let them out?

Vince: That is worse. Why would you want someone letting out mice?

Angela: Things happen. People have revenge issues.

Vince: But they are diseased mice. If you let out diseased mice it doesn't matter. Things like that happen. If there is a chance for someone to break in, what is to prevent someone from breaking in and letting out a whole bunch of other mice with worse disease? Aha. Take that.

Vince assumed that the mice were either released intentionally or unintentionally from a secure facility and this jeopardized its future security. Angela had a different view, assuming that someone may have staged the incident, and didn't want to let this influence her decision making. Generally players interpret this initial incident as they see fit, either as a critical flaw in the proposed facility if they are already inclined against it, or as a distraction if they are already in favor of the facility.

As they uncover the details and verify the credibility of the incident, it plays a less central role in players' arguments, and can even diminish the credibility of other NPCs or players who cite this information. Vince, from the exchange above, later stated:

I'm not sure whether it was a student or a professor but my impression is that it was stupid. How they got hold of the mice is shaky. Supposedly they have top security around the facilities and then someone can break in and steal the mice.

He went on to cite other information to defend his position against the facility. In written debriefs after the game, all but one or two students in most classes have come to believe that the incident is a hoax.

Promoting reflection and changes in rationales and ratings was the motivation for creating a more sophisticated scoring mechanism. POSIT's scoring mechanism was designed to encourage more frequent exchanges between players, and to force them to thoughtfully reconsider their opinions on additional occasions. Part of the scoring is supported by the software itself, and part is merely structuring the interactions between the players.

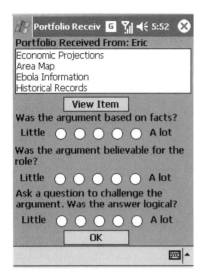

**Figure 11.8**
The rubric used when ranking incoming portfolios shows the three criteria used for
evaluating evidence.

We wanted the score in the game to be tightly linked to what we thought
should be valued in the game—making good arguments backed up with
evidence. To promote this, we created an in-game rubric that players used
to rank incoming portfolios (figure 11.8).

The primary criteria we wanted to make sure were present were rationales
based on *evidence*, presentations that *represented the role* well (as opposed to
the presenter's own opinion), and *ownership* of the evidence. The presenter
shows the evidence portfolio to the receiver who must then evaluate the
actual portfolio and the presentation (including a response to a challenge
question). The presenter then gets a score based on the receiver/evaluator's
rubric responses. Final score is determined based on the average score of a
person's portfolio, which has been evaluated by at least a minimum number of other players.

Players can exchange evidence portfolios with each other any time they
like. This can help them hone their arguments through feedback from the
other player, and any new evidence and insights gathered from talking to
them. But periodically throughout the game—the number of times can
be adjusted, though in practice it happened once every thirty to sixty
minutes—players were prompted to exchange evidence portfolios with

partners. In order to keep the scores honest, players were divided into two teams—Red and Blue. When players evaluated portfolios, they only evaluated portfolios from the opposite team, so that Red players evaluated Blue players' portfolios and vice versa. But players only competed with their team members for score; Thus there were two winners, the highest score from each team. This enabled evaluators to feel free to give scores without fear of hurting their own standing. While some players responded in the game summary that they felt their portfolios were evaluated unfairly, based on our own spot checks and subjective evaluations, this seemed to be associated with unreasonable expectations more than bias.

## POSIT Experiences

While POSIT was played by groups of adults, college students, and high school students, only the high school students had the use of the scoring rubric and team structure. These students were from an urban high school in the Boston metropolitan area. In many of the early trials without the scoring and team structures most people did not change their opinions over the course of the game. If we look at the magnitude of change in the high school students' opinions over the course of the game, we find that they changed on average two points. In this sample, about one-quarter of the student scores did not change (from start to finish, though they may have changed in the interim), while the rest changed at least one point and several of the students made dramatic shifts of five or six points, reflecting substantial opinion changes (figure 11.9). This sample is likely too small to make any general inferences, but students did change their opinions over the course of the game when these scoring mechanisms were in place.

As to what led the players to change their opinions, they stated that the their role was the single most influential factor, followed by the items found within the game (documents), and the news flashes that were presented to them periodically. Their own personal opinions were among the least important stated factors in changing their opinions along the way. At the same time, however, most of the players concurred that their own personal opinions resonated with those of their role. As the adage goes, "walk a mile in someone else's shoes."

Identifying with their roles was apparent in the dialogue that many of the students had over the course of the game. While one player was reading

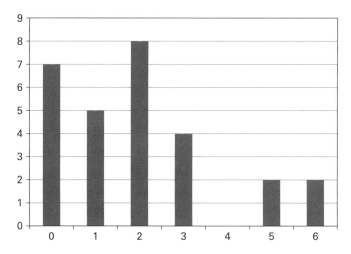

**Figure 11.9**
Changes in players' opinions (in absolute value) over the course of POSIT, showing many small changes and a few large changes.

an interview, one of the researchers stopped to ask her what she was doing. She replied: "I'm talking to some guy named "Felix." He works with infectious diseases. He thinks that it would be good to prepare for bioterrorism. So he is for the BSL-4 lab. He works in a BSL-3 lab and thinks it is incredibly safe. That is one of the things I was worried about. I didn't exactly know the different levels of the laboratories. Now I know 1 through 4 where 1 is the safest. I work at a BSL-1 lab."

This short excerpt shows the player moving seamlessly between her own knowledge and her role's perspective. She is actually the one "talking" to the character, though her role is the one who works in a BSL-4 lab. As the player, she didn't know the differences between the levels of the biosafety labs, and it isn't clear whether she or her character was worried about the lab safety, given how quickly she moves between talking through both perspectives here.

Other players stated that they tried hard to use a lens or filter for all of the evidence that would be appropriate to their role, but depending on how extreme that role is they had a hard time finding enough evidence through that filter. "I tried my best to stay in the −5 region but found it hard to find supporting evidence," said one player, "and therefore I neglected anything positive about the facility."

The players whose roles presumably had less extreme or more flexible opinions, however, were able to successfully employ that filter. "My role's opinion changed because I made the assumption that a local parent would not want this building in his/her neighborhood," said one player. "Based on that assumption I looked for evidence that was in line with it [a negative opinion]."

Good evidence-based arguments should combine both a strong basis in fact as well as appropriate emotion. These uses were modeled in the dialogue of the NPCs and players learned the effectiveness of these arguments from experience. Over time players were able to integrate these same two components into their own arguments. For example:

Tonya:   [If something bad happens] I might die or get sick or something and might not be able to see my husband or my children ever again. But it would also be good because it provides jobs for construction workers or new biologists who would want to work in a BSL-4 lab.

Vince:   Okay. Let me look at your score [current rating]. I think you had a lot of good things to say. Especially that [interview] because it was based on fact but it also gave a person's perspective on it. I wasn't too sure about the first one [piece of evidence]— I was a bit confused and led astray about all of the monkeys.

In this case, Tonya is mixing an emotional argument about what might happen to her, and the impact that would have on her husband and children, with some concrete arguments on job creation as a result of the facility being built. Vince clearly appreciates that balance in her presentation, and explicitly states that he has valued those two components in one of the particularly convincing interviews that she had presented in her evidence portfolio.

The scoring system was viewed favorably by players who stated that it made them think more about how to support their arguments. Players stated that the scoring system

• "made you realize some things did not back your argument as much as you thought."

• "helped you to make your argument stronger and efficient."

• "made people have to back up their ideas."

In essence, the scoring system accomplished exactly what it set out to do, and built the students' skills in constructing and evaluating evidence-based arguments.

## Place Basis

Many of the students complained about having to walk great distances as a part of the game. This was clearly exacerbated by the weather, which was quite cold and wet over the course of these runs. Even though the game takes place primarily indoors, players did spend time outside near buildings and in between buildings. The students likely would have preferred placing all of the NPCs and evidence in one room, and having immediate access to that information.

The current arrangement, however, forced the players to think carefully about what evidence they needed. They could not get all of the information from everyone, but rather needed to plan a strategy for getting the data they thought would be most effective. Clearly some of the evidence gathering was opportunistic, as players would interview characters they encountered were along the way. But access to critical pieces of evidence needed to be planned.

Further, as in the other AR games, the location mattered. This game was not about a site that was far away or fictional. It was about the neighborhood, community, and campus the players were in. The crux of the issue in this scenario was location-specific. Most of the people would not mind if a BSL-4 lab were built somewhere, but it mattered that it could be built in their own backyards. As one player stated, who changed his opinion from initially positive to negative in the end, his opinion changed "because I saw the spot where the building was set to be built on and it was very scary how many students and people walked by it constantly."

This suggests that there are two benefits to distributing POSIT information spatially. First, it provides for a means to reward planning and reflection on information collection. For this reason, POSIT games could be played anywhere, including campuses, museums, and workplaces. The second benefit is collecting additional tacit information that pertains to the question at hand. This benefit is less universally applicable, but could be used as a tool for many decisions that need to be made in particular locations, ranging from rerouting of roads, to construction of new facilities, to changes in land use. These are controversial and complex decisions that frequently get made across a range of spatial scales. POSIT is a way for stakeholders to grapple with these issues and walk a mile in someone else's virtual shoes.

# 12 Anytime, Anywhere: Palmagotchi

Internet legend (spread on popular sites such as ⟨tamatalk.com⟩) has it that in 1996, a Japanese housewife named Aki Maita noticed that Japanese teenage girls had a lot of idle time sitting on trains going back and forth to school. She thought that those girls would love a fun little toy to occupy their time. Maita's inspiration was to couple people's interests in pets and gadgets in her idea for a small electronic pet one could carry around on a key chain. She told this idea to her friends who encouraged her to pitch it to toy giant Bandai. After some prodding she did pitch the idea, and Bandai bought it and created the Tamagotchi, an egg-shaped electronic virtual pet that went on to sell millions.

That is an interesting story, but only slightly more interesting than the actual origins of the famous virtual pet. According to more reliable sources (Wudunn 1997; Time Asia 1998; Beech 2000) Aki Maita was a low-level supervisor working at Bandai, who cocreated Tamagotchi with Akihiro Yokoi, a former Bandai employee and executive at the Wiz Company, a toy design company. Yokoi was also a pet lover who though it would be great to capitalize on people's love for pets along with the convenience of small electronic gadgetry. Together they designed the device, with the aid of other toy designers at Bandai and Wiz. Maita and Yokoi, along with their respective companies Bandai and Wiz, hold most of the patents on Tamagotchi jointly, and in 1997 they received the infamous Ig Noble award for Economics (Improbable Research 2007) as the mother and father of the Tamagotchi, "for diverting millions of person-hours of work into the husbandry of virtual pets."

From there the story becomes a bit more clear. The Tamagotchi, so named because of the egg shape of the toy (*tamago* is "egg" in Japanese, combined with the affectionate suffix *tchi* meaning "cute," and is also a play on the word *Tamadachi*, meaning "friend"), was a runaway success in its early days. Released late in 1996 in Japan and early in 1997 worldwide, the first run of the device sold out quickly, so quickly that in Japan Bandai employees had to remove company identification from their apparel for fear of theft.

The initial version of the Tamagotchi in 1996 took the form of a small colorful egg with a miniscule black and white LCD screen and three buttons. The game starts with the birth of a digital creature that the owner of the Tamagotchi must feed, train, discipline, and clean up after using the simple three-button interface. A well-cared-for Tamagotchi grows through several life stages from toddler to teen to adult to grandparent and develops positive attributes. A poorly cared-for Tamagotchi will develop undesirable characteristics and either die (in the Asian versions) or fly off (in the North American versions), forcing the owner to start again with a new Tamagotchi.

After the sellout first run, Bandai ramped up for another run (version 2) of the Tamagotchi in 1997 and 1998. Despite the runaway success of the Tamagotchi, Bandai faltered in these early years, partially due to the unexpected rapid growth of its product, and partially due to the arrival of Furby in 1998, which dethroned the Tamagotchi as the must-have toy. Bandai wound up with a surplus of Tamagotchis unsold, which caused significant disruptions to the company at the time, despite sales of forty million units (Reuters 2006). While many knockoffs lived on for years, including notables like NanoPets and GigaPets, the Tamagotchi ceased production in 1998, and lay dormant until 2004 when the Tamgotchi Connections (version 3) line was introduced to much fanfare. Version 3 built incrementally on the Tamagotchi model, offering a nearly identical shell, LCD screen, and three-button interface. Version 3 notably introduced additional minigames used to train the Tamagotchi, as well as peer-to-peer interactions via an infrared port and an online connection that could be accessed through passwords provided to the player by their Tamagotchi. While some may see the peer-to-peer interactions as merely a marketing ploy to prod players into getting their friends to also purchase Tamagotchis, this feature opened up a new realm of social game play. Players could compete with other Tamagotchis, socialize with other Tamagotchis, and mate with other Tamagotchis (mating results in twins colonizing each of the players' Tamagotchis). This interactivity perhaps borrows from the successful de-

**Figure 12.1**
A version 4 Tamagotchi Connection near birth. The three-button interface, key-chain, and simple black-and-white LCD screen have been consistent design features across multiple versions. Available at ⟨http://www.tamagotchi.com/products/catalog1.php⟩.

sign of the Lovegety, a popular pocket-sized followup to the Tamagotchi launched in 1998. The Lovegety could wirelessly search for other nearby Lovegetys of the opposite gender, leading to face-to-face interaction. Building on this peer-to-peer model, Version 3 sold a respectable twenty million units in its first two years, leading to the introduction of Version 4 in late 2006 and early 2007. Even Version 4 holds true to the original Tamagotchi image, strengthening the online connection and further individualizing the Tamagtochis through role-playing of professions that the Tamagotchis can pursue.

While the Tamagotchi went on hiatus from the late 1990s until 2004, the virtual pet business thrived more broadly. The behemoth in the industry is Neopets, a company that according to its Web site was founded in 1999 by Adam Powell and Donna Williams, two British college students. The free online virtual pets site, bought in 2005 by Viacom, claims to have over seventy million users. Like Tamagotchis, Neopets require feeding and care, but there is a much greater emphasis on side games, battles, and social networking. Years later, in 2004 (in Japan; 2005 in North America),

Nintendogs hit Nintendo's DS platform, bringing a lifelike dog simulator to a portable platform (see chapter 4 for more details). Nintendogs capitalized on the built-in wireless networking, microphone, and touch-screen of the DS to make the simulation quite realistic.

Another breakout success in the virtual pet industry came through the innovation of Webkinz, a Ganz company. Webkinz married the virtual pet sensation of Tamagotchi with the stuffed animal sensation long popular with young girls. Webkinz is an online virtual pet community, but the only way that one can get an account is by buying a stuffed animal "Webkinz" that comes with account information. The online account is attached to the physical stuffed animal, creating an even greater bond for players (typically young girls age ten and under). While I personally haven't taken the Webkinz plunge like I have with Tamagotchi and Nintendogs, I have observed countless hours of play by my niece. As a prekindergartener she has been known to play three to four hours a day. I would doubt if her parents (or many parents of prekindergarteners) would buy a subscription to an online game. That isn't a problem for Webkinz, since you can't buy (or even renew) an account. The only way to play is to buy (and keep buying) the stuffed animals, each of which has a one-year subscription, and parents have a long history of buying their kids stuffed animals.

In the Webkinz world, like many of the other online virtual pet communities, owners need to care for their pets through feeding and maintenance, but also can buy items for their pets to spruce them up and decorate their homes. Children earn points or money towards those purchases through solitary or head-to-head game play with other Webkinz owners. In fact, most of the time spent in the world of Webkinz is playing these casual games. The actual actions of caring for one's Webkinz pet (e.g., feeding it) take almost no time, but earning the points required to buy your Webkinz its necessities through game play does take substantial amounts of time. Interestingly, Webkinz maintains its child safety through structured communication between players that can only be chosen from predefined dialogue sequences.

Many kids (my niece included) will maintain multiple pets at a time, which requires spending quite a bit of time in the world of Webkinz. Some critics would be quick to point out that such games promote excessive screen time, well beyond the hour or so a day that many of them promote as a hard limit. These limits, however well intended, are not based on substantial evidence, particularly in light of an evolving digital society and our relationship between the physical and digital worlds. I don't want to say that there shouldn't be limits, as the question needs further study, but I can say that I have observed many positive outcomes from Webkinz play. I have seen creative digital play, exploration of roles and characters, and deep learning about a complex system. My niece has learned how to balance care for multiple pets, what to play for quick points to feed them, and where to go to socialize. At the same time I have observed the behavior of other players in the world of Webkinz, such as the other Webkinz who quickly left the spelling bee after my niece handed over the game to her father to play. His concern for the health of his daughter's Webkinz, while a bit of an obsession, has actually become a social activity through which a father and daughter can bond.

Hundreds of virtual pet sites have emerged over the years. The Web site ⟨virtualpet.com⟩ provides a catalogue of some of these, listing hundreds of sites caring for pets ranging from dogs to dinosaurs. However, Tamagotchi, and to some extent Nintendogs, stand apart from these games in some fundamental ways. Sites like Neopets, Webkinz, and others use the virtual pet as an anchor of much broader game play. Social interaction, battle, shopping, and a host of mostly irrelevant casual game play are the real focus of these sites. Pet maintenance and care takes little time relative to these other activities, and is not the primary focus of players' visits. Engaging in these related tasks necessarily takes on greater sophistication and complexity. In contrast, the Tamagotchi, with its three-button interface and black-and-white screen is a statement in simplicity. The focus of Tamagotchi is the maintenance and care of your pet. While there are short games that one can play in Tamagotchi, for the most part they are only of interest in serving the wellbeing of your pet. There is little opportunity to play Tamagotchi for hours a day. Though a player may think about their Tamagotchi all day, screen time is typically measured in minutes a day. Unlike onscreen worlds in which the player is only deeply engaged when they feel immersed in this virtual world, the Tamagotchi player remains in our world at all times, though the digital pet becomes integrated in the life of its owner. This distinction in simplicity, goals, and integration versus immersion distinguishes Tamagotchi from the vast majority of contenders in the virtual pets field, as well as those in The Sims genre or games like Harvest Moon that also involve maintenance of creatures.

## Kids, Teachers, and Tamagotchis

The virtual pet recipe seems simple enough. People form deep emotional attachments to their pets, and the digital pet capitalizes on that inherent connection and puts it on screen. However, on first glance a Tamagotchi displays none of the typical characteristics associated with animals that help create this bond. Tamagotchis don't have big eyes. They aren't warm and cuddly. They don't come when you call, keep away the mice or fetch the paper (though the same could be said for many of my actual pets). The Tamagotchi's tiny black-and-white LCD and its simple beeps are the utmost in minimalist design and representation. Yet people deeply connect with their digital pets, and only play because they care what happens to them.

Sherry Turkle (1995, 2005) has studied the relationship between people and many forms of digital representations, including Tamagotchis and the robotic Furbies. She has found that kids have begun to recognize these kinds of digital beings as "sort of alive." While kids understand that a digital creature is *alive* in a fundamentally different way than an actual pet, they ascribe a different form of *being alive* to digital creatures. Hence, they care about their digital pets, tend to their needs, and are invested in their outcomes because they are *alive*. Kids look past much of the physical representation to form a relationship with this digital artifact. Referring to Tamagotchis and Furbies, Turkle (2005) says, "In the case of the toys, the culture is being presented with computational objects that elicit emotional response and that evoke a sense of relationship... there is less a concern with whether these objects 'really' know or feel and an increasing sense of connection with them. In sum, we are creating objects that push our evolutionary buttons to respond to interactivity by experiencing ourselves as with a kindred 'other.'" (p. 277)

The simplicity of the design of Tamagotchi, along with this emotional bond and associated investment that it creates, provides a perfect model for educational games. It is a game that is simple to learn and interact with, creates an attachment to the outcomes for the players, allows for building of mastery, can be played casually in bursts of a few minutes at a time, and yet sustains interest and interaction over long periods of time. Of course, the Tamagotchi's relationship with schools has not necessarily been positive in the past. As with many successful portable toys, the Tamagotchi plagued many teachers in its early run in the late 1990s, leading schools around the world to ban virtual pets. Bandai quickly heard the message and included a pause feature in version 2 that allowed kids to give their Tamagotchi (and their teachers) a break from all the beeping, thus making the game a bit more school-friendly.

## The Birth of Palmagotchi

In many ways our Palm participatory simulations model the design of the Tamagotchi. Their simplicity, attachment to outcomes, and social interaction are shared with the Tamagotchi. However, all of the Palm participatory simulations were designed to take place *in class*, not in the hallways, cafeterias, or libraries occupying the "free" time throughout many students' school days. Thus, as they were, the participatory simulations faced the fierce competition for in-class time that most technologies confront, despite their easier access and fit into the classroom model. We thought that we could advance the PSims model by bringing game play out of the classroom while tying outcomes and learning explicitly to in-class content and discussion.

Thus the idea of *Palmagotchi* (cute Palm) was born, a simple educational game played occasionally over long periods of time outside of class, connecting back to classroom content through discussion and data analysis. Many of the virtual pet communities contained the essential elements of learning games, requiring players to understand their organisms' requirements for survival and even mating. Building on this idea, Palmagotchi would be a game that drew upon analogies to Darwin's finches in the Galapagos. Players would have to maintain families of birds and islands of flowers all the while considering issues in genetics, ecology, and evolution to best maximize their chances for survival.

The name was perfect. Unfortunately, the platform turned out to be less than perfect. Palm was a great platform for the simple interactions of the participatory simulations, but more complex transactions and data management proved to be easier on the Pocket PC (Windows Mobile) platform. Nonetheless, the name stuck as we were not particularly fond of PocketPCagotchi or WindowsMobileagotchi.

While we were successful in maintaining simple game play, development was complex and the game took years to create. Much of that complexity

**Figure 12.2**
The diversity of species in the finches of the Galapagos is easily seen in their beak morphology. Some beaks are broad and stout for crunching on seeds and nuts, while others are thin and narrow for foraging on flowers. (From Darwin's original illustrations, see Darwin, C. R. 1890. *Journal of Researches into the Natural History and Geology of the Various Countries Visited by H.M.S. Beagle etc.* London: John Murray, first Murray illustrated edition.)

originated from the requirement that the game be client-server-oriented. This requirement emerged for two reasons. First, we wanted to be able to collect data in real-time and use that as a point of discussion in classes. Second, given that the game would be played continuously, we wanted to remove the requisite face-to-face interactions, permitting interaction from afar. This would allow players to interact with their friends even when the friends weren't nearby.

The first complete version fairly closely modeled the scenario of Darwin's finches in the Galapagos. For those not familiar with the details of Darwin's finches, the history is as follows: Charles Darwin collected a number of samples of seemingly very diverse bird species on his famous voyage of the *Beagle* to the Galapagos Islands in 1835. Upon his return to London he discussed these samples with other scientists, and later came to find that these species were not as unrelated as he previously thought. Instead, these species all diverged from a common ancestor species of finch that arrived in the Galapagos Islands from the mainland of Ecuador. After arriving on one island, the species spread out to other islands. Each of these islands had unique food availability due to diverse habitats and other species present. Those finches that were best suited to surviving on their island's particular food sources, primarily due to differences in beak morphology, thrived and reproduced. Over time, due to separation and strong selection for these differences, new species emerged that appeared dramatically different from one another.

This canonical example of evolution is told in countless classrooms around the world, yet the dynamic interactive nature of evolution remains poorly understood by most people, particularly Americans. Evolution must be understood as a process that takes place over time, takes place at the population level, and is influenced by random events (i.e., things don't have purpose in evolution, despite the pretext of some science fiction and the teachings of many "alternative" theories). Palmagotchi embodies these features by creating a world in which each player maintains an island full of flowers and a family of birds. The actions that one may take in Palmagotchi are simple. Birds can feed from flowers (on other players' islands, just to keep it interesting), they may mate with birds from other families, and they may feed their offspring as they grow. However, these few actions involve a lot of decision making.

## Game Play

All of the birds and flowers in Palmagotchi are endowed with a number of genetically determined traits. Each trait is additively influenced by a small number of copies of the same gene. This means traits are not simply present or absent (e.g., red or white flowers) but rather are continuous, spanning a whole spectrum (e.g., height of a flower). Flowers have the following genetic characteristics:

Color   The color of the flower, which influences which birds forage from them.

Flower length   The length of the flower, which influences which bird beak lengths can forage from them. Better matches of beak length and flower length result in more nectar for the bird.

Pollen type   The texture of pollen, which influences which birds will carry it when foraged upon.

Production rate   The rate at which flowers produce nectar and pollen.

Heartiness   A characteristic that influences the chance that flowers will be killed or damaged during cold spells.

Birds have a complementary set of genetic characteristics:

Size   The mass of the bird, which affects the bird's speed.

Metabolism   The rate of energy burn for the bird, which in turn interacts with speed and ability to survive in the cold.

Speed   The speed that birds fly, which makes the bird less likely to be attacked while foraging.

Feather type   The quality of the feathers that determines how sticky they are for the pollen.

Featheriness   A measure of quantity of feathers that makes a bird more resilient to cold spells, but decreases its speed.

Beak length   The length of the bird's beak. Better matches of beak length and flower length result in more pollen and nectar during foraging.

Color preference   Determines what colors the birds can see, which in turn affects which flowers a bird may forage from.

These characteristics are determined entirely genetically; that is, there is no variation contributed by the environment. But there are dynamic traits that do change over time. Dynamic traits are influenced by a combination of time, random events, genetic traits, and the input and actions of players. For flowers these traits are:

Pollen   The amount of pollen that birds may pick up from a flower.

Nectar   The amount of nectar that birds may forage from a flower.

And for birds they are:

Age   A measure of time that the bird has lived. Birds are designated as juveniles until age five. The timescale can be arbitrarily mapped to real time.

Energy   A bird's food reserves. This quantity is depleted by mating and by the bird's metabolism. Birds gain more energy by foraging for nectar.

Players access information about their birds and flowers in two ways. They have quick access via their handhelds to some of the important characteristics at a glance on the main overview onscreen page for their birds and flowers. These pages display a list of all birds and flowers that the player currently has on their island, along with a picture of each bird or flower (which is determined via a simple algorithm), bar graphs of key characteristics, and a randomly generated name. By clicking on any one of these birds or flowers the player accesses detailed information about the genetic

**Figure 12.3**

Overviews of the birds and flowers (a and c) showing pictures and basic statistics on each bird and flower, along with the detailed view (b and d) showing all of the genetic and dynamic characteristics.

characteristics and dynamic traits. The analogy we make is that the overview page shows the front view of a trading card, and once the player clicks on a bird or flower it gives the back view with details and statistics.

Players begin Palmagotchi by creating an ID and wirelessly logging into a central server. This facilitates all future interactions with other players. They are then presented with their initial set of birds and flowers. The game is paced to require interactions on the course of every one to three hours so as not to disrupt classes, yet create some sense that the players must be vigilant to keep their organisms alive and well.

Each interaction is designed to present the player with data that they can use to inform their decisions, though the only way that the player learns how this data maps on to success is through experience. For example, a player looks across their current set of birds and decides which one needs to forage. After selecting that bird, they select from a list of online players' islands that they can visit. Once on that island the player is presented with a list of flowers that they are able to "see" (figure 12.4a). Each flower has a color, and each bird has a color preference. The game automatically filters

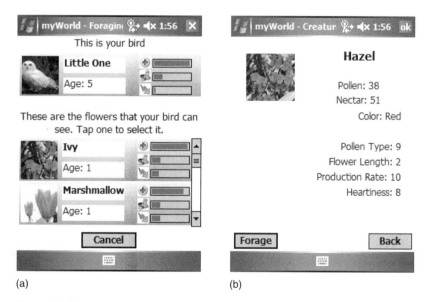

(a)                                          (b)

**Figure 12.4**
The list of flowers that the foraging bird "Little One" can see (a) provides the player with a quick overview of available resources on the chosen island. Further examination (b) can reveal individual characteristics of each flower.

out those flowers that are too much of a mismatch between the colors the bird can perceive and the color of the flower. Looking at those preferences as points on the spectrum, birds can see flowers that are within a given distance from their preferences on that spectrum. At a glance the player can see the visible flowers and their current health. By clicking on any one of those flowers the player can further examine each of the genetic characteristics of those flowers (pollen type, flower length, etc.) as shown in figure 12.4b.

In the early stages players may have little reason to choose among the particular characteristics, but soon they learn that there are several consequences to their foraging actions and may begin to alter their choices in light of those consequences. First, there are *direct impacts on the foraging birds*. Birds can gather more nectar from foraging when their beak length closely matches the flower length. In most cases the birds will be able to get some nectar regardless of the match, but they may do well to minimize trips and maximize the nectar taken in any one trip. Players soon discover that there is a risk associated with each foraging venture, for the birds may be attacked by predators during each foraging attempt (figure 12.5). Players

**Figure 12.5**
An alert given to players when they are attacked during foraging lets them know of the risks associated with each foraging venture.

learn that these risks can be mitigated by reducing the total number of trips the birds need to take in addition to selecting for birds with increased speed. Metabolism turns out to be a double-edged sword, players find. Increasing metabolism correlates with increased speed, but also burns resources more quickly, resulting in a need for more foraging trips. This complex relationship between foraging behaviors and the characteristics of birds and flowers is one of the interesting challenges that the Palma-gotchi game presents.

There is also an *indirect effect of foraging on the traits of the flowers.* When birds forage they pick up pollen on their feathers. Some of that pollen is transferred to the next flower that they forage upon. If the pollen on the two flowers is a close enough match, they can produce offspring that combine the genetic characteristics of the two parents. Thus foraging from two similar flowers in a row will likely produce more of that flower type, while foraging from two flowers with quite different characteristics will likely result in no offspring, or offspring that express characteristics that are a hybrid of the two.

In addition to foraging, birds must also mate. Even if well tended to, the birds have finite lifespans; thus mating and producing offspring are a regular part of the game. The in-game actions required for mating are much like those for foraging. Players select the other player (island) they want to visit. In this case, since there is some sense that both birds may be selective, the chosen partner has the opportunity to decline the visit to their island. If they accept, the first player is presented with a list of birds with whom theirs can mate (figure 12.6a). They can see the health overviews of those birds immediately, and then delve deeper into their information by viewing the birds' genetic statistics (figure 12.6b). After choosing a mate, each of the two players is given a new set of juvenile birds. The number that each player receives is dependent on the "clutch size" characteristic of the parents. It is also important to note that after this mating, any other adult or juvenile birds that the player had, with the exception of the chosen parent, disappear from their screen, leaving only the parent and the new offspring (figure 6c).

Players may initially mate their birds out of convenience with their friends or other players who are available. But over time selection for desired characteristics becomes a key strategy for survival. Players may determine that increased speed is desirable to avoid predators, or that smaller

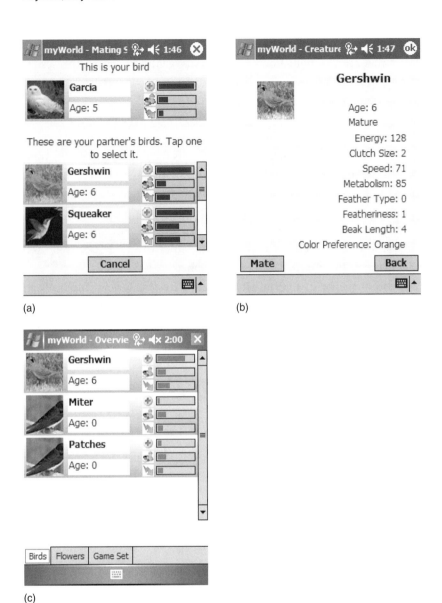

**Figure 12.6a–c**

The process of mating involves selecting a bird to mate and then finding a partner (a) with the desired characteristics (b) on another player's island. After a successful mating, the parent remains, but other birds disappear, leaving only the parent and the new offspring (c).

beak length is desirable to feed from available flowers. Offspring possess genes that are a combination of the parents' genes. In general this results in offspring that are near the average of the parents' two traits, but due to the random assortment of genes, not all offspring will be exactly the same.

After juveniles are born, they must be fed by the remaining parent until they mature at age five. In order to feed offspring, the parent must first obtain nectar by foraging. Then the player selects the offspring that they want to feed, and distributes that nectar to the offspring using a sliding scale. The player may choose to transfer all of the nectar to an individual offspring, or distribute that quantity to several different offspring (figure 12.7). This is another decision point for players. If they have large clutch sizes they may choose to distribute energy evenly among all of them, or selectively feed the ones that they think have the best chance for survival.

Sometimes, due to unfortunate events or bad strategies employed by players, all of a player's birds or flowers may die. At that point the player's island can be recolonized by birds or flowers from another player's island. A player can select one of their birds or flowers, which will leave their ma-

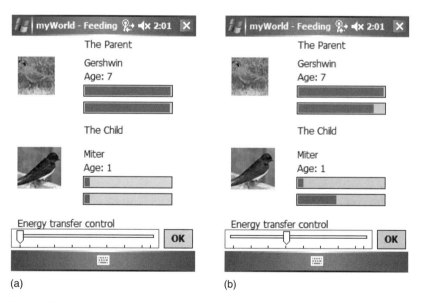

(a)                                            (b)

**Figure 12.7a–b**
The screens showing how players feed their offspring, both before (a) and after (b) nectar has been transferred. The parent bird can choose how much nectar to transfer.

chine (island) and appear on the other player's handheld. It is important that this is merely redistributing birds and flowers already in the population, which doesn't affect the genetic diversity of the community.

## Learning about Evolution

Despite some of the strongest evidence among scientific theories, the majority of Americans reject the theory of evolution. A 2005 Harris Poll (Harris Poll June 17–21) showed that only 46 percent of people surveyed believed that Darwin's theory of evolution was supported by fossil records, and a mere 38 percent believed that human beings evolved from other species. Other polls have found similar results, including an annual CBS News poll that has consistently found more than 50 percent (51–55 percent from 2004–2006) of people polled believe that humans were created in their current form, and just slightly below 50 percent (44–48 percent from 2005–2006) believe that humans were created in the last 10,000 years.

While some of these views are supported by fundamental misunderstanding or misuse of scientific methodology and evidence, additional misconceptions persist even among believers in evolution (Engel Clough and Wood-Robinson 1985; Bishop and Anderson 1990; Alters and Alters 2001). Evolution, simply put, is a change in the genetic frequency of characteristics over time caused by genetic drift (random fluctuations in gene frequencies), gene flow (immigration and emigration), and natural selection (Darwin's "survival of the fittest"). The misconceptions that people hold are often due to misunderstanding the mechanisms by which these processes work, hence the goal of Palmagotchi to have people "see" evolution in action. Specifically, Palmagotchi was designed to help people understand that evolution is a dynamic process influenced by both random events (e.g., chance events in the environment) as well as the environment (e.g., availability of food, current climate, etc.). At the same time we wanted players to learn the relative importance of factors affecting evolution, including the relatively rare occurrence of mutation (which doesn't happen at all in Palmagotchi) and the more common events affecting survival and reproduction of individuals.

We did need to be careful in the design not to introduce or reinforce notions about evolution having a "goal," since players do influence the direction of evolution through their choices. We mitigate this by directly addressing that issue in conversation, and by trying to keep the actions within the game similar in scope to the behavior of birds. While birds may not be able to plan for future generations, they can make decisions based on colors and shapes of foods, and even distinguish some "suitability" characteristics in mates.

We found that students did indeed develop a deeper understanding of evolution through game play. Our first run of Palmagotchi was with a high school senior class in biology, which played the game over a three-day period. Instruments assessing students' understanding of evolution were given at the beginning of the activity and upon completion of game play. For example, on one of the questions asking whether evolution was predictable, one student replied, before the activity took place, "Usually, because you can tell whether a species is more fit than another, but sometimes mutations can make this unpredictable."

This answer expresses a number of misconceptions. First, it confuses evolution happening at a species level (incorrect) versus the population level. Second, it mentions "fitness" of a species absolutely (incorrect), instead of fitness of individuals relative to the environment. Finally, it promotes mutations as being a primary source of unpredictability, when in fact while these are important contributors they are extremely rare. After the game the same student responded to the same question on whether evolution was predictable: "Not really because…natural disasters like cold spells, etc. are unpredictable. Pretty much birds with the "best" characteristics would survive, but what is considered the best can change depending on what type of environment you are in."

After the game, the student no longer referred to species or mutations. Now her answer focuses on some of the unpredictability inherent in the environment, and the fitness of individuals relative to the environment. While this radical improvement in understanding was at the higher end among students in the class, it does show that the game could significantly improve understanding.

But other improvements were seen. Another student initially responded to a question about the relationship between genetics and evolution by saying, "Changes in natural genetics cause evolution. Mutation of genes and proteins is the cause of evolution."

This student identified a link between genetics and evolution, but placed the burden of this link on mutation (including the incorrect "mutations" of proteins). After the game the student more concisely stated: "Genetics cause evolution in relation with the environment."

While this game didn't help students articulate strong written answers, it did eliminate the emphasis on mutation, and strongly showed the influence of the environment. Several other students also emphasized mutation initially, and later either qualified or diminished that importance and more thoroughly included the relationship of the individual and population with the environment.

After asking students to individually reflect on their answers, the class turned to a group discussion to investigate how their population of birds and flowers changed over time. Students were asked to plot out the distribution of certain characteristics to investigate whether evolution did in fact occur in their population. For example, in a plot of the beak length of the population, the population was somewhat randomly and evenly distributed at the beginning (figure 12.8a), but had a much greater central tendency in the end (figure 12.8b). While some of this shift in frequency may be related to statistical phenomena, it does show a change in the population over time.

In a discussion of the plots, students were asked what happened, and why there might be this central tendency. One of the students responded: "We selected for the middle [beak length] of the road. Since if we had a four, you could eat from [flower lengths] five, six, or seven, as well as one, two, or three. But if you had a [beak length] seven then we couldn't."

This is known as stabilizing selection. The students selected for the middle traits since they were better adapted to the environment (able to feed essentially from all available flowers). Individuals with the extreme traits (very large or small) were confined to a smaller pool of resources. Given that there was even availability across this distribution of flower lengths, the resulting distribution of beak lengths matches well with what one should be able to predict.

Similarly, while this was not witnessed in the course of the game, students predicted that in the long run we should see a trend towards flower lengths four and five. Said one, "Everyone is choosing beak length four or five, so everyone will eat from the flower lengths four and five [since there is the closest match], and then flower lengths seven to nine will die out."

This prediction makes a lot of sense given the data that was collected, and it recognizes the scale of evolution both in size (populations), and the

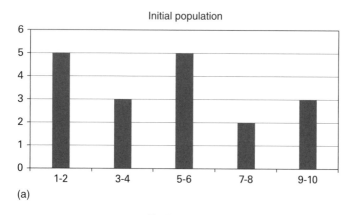

(a)

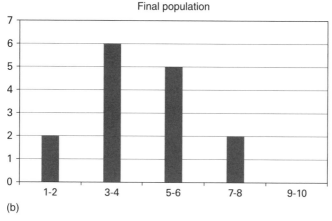

(b)

**Figure 12.8a–b**

Plots of the beak length of birds in Palmagotchi at the beginning (a) and end (b) of the game, showing a much greater central tendency at the end.

dynamic feedback required for such changes to occur. However, birds with beak length four or five can (and do) still eat from other flower lengths when they were available, slowing the trend towards mid-length flowers. As one student said: "If there is only one flower that is the only option that you have. You can't always control what you're going for [due to small samples]. So it is going to vary how fast it [evolution] can go and maintain diversity." This is recognition of the time scale over which evolution occurs and the dynamic forces at play maintaining diversity (flattening the distribution) and selecting for narrow ranges of traits (narrowing the distribution).

Students similarly recognized the course and timescale of other traits. Some selected for large clutch sizes to "increase the chances that some of the offspring will be of good quality." Others considered the effects of speed and metabolism on their birds' ability to survive attacks by predators. While cold spells are not particularly frequent, they have noticeable effects early in the game. As a result, many students selected for birds with more feathers and higher metabolism. Since these traits have costs, students may better estimate the probability of cold spells over time and reconsider those choices, but over the course of this game the traits persisted despite those costs.

In another class playing Palmagotchi over a one-week period, early in the game there was a technical glitch that caused the birds on all but two or three handhelds to die. In order for the other players to continue, they needed to be recolonized from the existing stock of birds on the surviving handhelds. This meant that the genetic diversity was limited to those surviving on the small number of handhelds, akin to a "genetic bottleneck" that happens in species when there are only a few individuals of that species left. This bottleneck became evident in the final analysis of the game when players noted that while they had originally started with a diversity of flower colors (every color in the rainbow), the vast majority at the end were green. Connected to this was the fact that the birds had a random distribution of flower color preferences at the beginning, and at the end they had a large majority of green color preferences. Presumably, by chance, the preponderance of surviving birds at the time of the bottleneck had a green color preference. This in turn selected for green flowers, which had a selective advantage given the surviving population of birds. Thus a great "natural" experiment in selection became evident in this population.

## Palmagotchi Game Play

Perhaps the greatest reason for the success of the Tamagotchis is that players were attached to their virtual pets. They cared what happened to them, and attached themselves to these simple digital creatures. This same investment was apparent in Palmagotchi. Players genuinely cared what happened to their creatures, despite the fact that their participation in the game was anonymous (from a research perspective) and did not count for their grade in class. It came down to attachment to creatures. One boy in the class said point blank: "I got very attached to my birds and I got very addicted to the game. So I was playing whenever possible—in between classes and free periods. I was playing during lunch."

While the use of the word "addiction" may set off some alarms, the behavior demonstrated is far from actual addiction. The amount of game play still amounted to a mere number of minutes over the course of the day. But the player did feel a deep emotional attachment to his birds and cared about their fate. Another boy from the class echoed that by saying: "It [the game] was extremely stressful. I was really concerned for my birds. When no one was paying attention I was like, 'please let me forage!'"

In general students felt quite attached to their birds, and wanted them to survive. Surveys showed that students agree fairly strongly about their responsibility for their birds (on average 3.27 on a 4-point scale). This responsibility measure is a good indicator of the investment students made in their creatures. In order for students to want to learn and develop better strategies (and learn the associated science that goes with those strategies), they need this attachment to their creatures and their outcomes. The success of Palmagotchi is rooted in motivating students to care about their virtual pets enough to thoughtfully consider the mechanisms at work and modify their strategies to reflect that understanding.

While the attachment of students to their creatures was good news, the student's comment preceding, however, raises an issue that came up over the course of the game. The class that played Palmagotchi was a small one, fewer than twenty students. When trying to play quickly between classes they needed to rely on other students turning on their PDAs to play during those same few minutes. If a player tried to forage or mate, there needed to be another player to accept that action. As a result, many of the interactions were facilitated by face-to-face play, even though line of sight was not required from a technical perspective. Students said that they "interacted with whoever. Not just my friends, unless my friends were who I could track down." Other players confirmed the value of face-to-face time, and several cited the few opportunities when everyone happened to be in the library or at lunch as popular times to play. On average the students reported 83 percent of their interactions were face to face as opposed to remote, with no student reporting less than 75 percent of their interactions taking place face to face. This suggests that future iterations should be played with larger groups, or include the option for asynchronous game play. One could argue, though, that the face-to-face communication requirement was an asset and not a liability. The face-to-face component can facilitate additional communication and conversation that may also be germane to learning. It also permits flexible communication and tangential conversation and interaction, which may well have added value. For example, players might ask each other why they chose a particular bird to mate, share recent stories about their virtual creatures, or make reference to their strategies in passing.

Again, like Tamagotchi, game play tended to be frequent but for short periods of time. Players described the average day as consisting of six to ten bouts of game play, each lasting anywhere from a few seconds to a few minutes. This time was enjoyed by the students (2.91 on a 4-point scale of enjoyment with 4 being the maximum). Students indicated that they were playing the game in total somewhere in the range of fifteen to twenty minutes per day, yet they described the experience as one in which they often thought about their digital creatures, even if they weren't looking directly at them. The big picture here is that in a relatively short amount of time each day the players became invested enough in their birds to engage deeply in the game. Many of the players indicated that this attachment and investment would have grown in subsequent days as they were able to devise better strategies.

## Palmagotchi Evolution

The first iterations of Palmagotchi were just being completed as this book is being written. As a first-generation product, there is a lot to be learned from early implementations, and many changes to come in the future. Players suggested additional features (e.g., the ability to name birds) that would increase their attachment to their creatures, and enhanced capabilities for the aforementioned asynchronous interactions. They also notably mentioned the inclusion of mini-games. Instead of merely selecting the flowers upon which the bird needs to forage, for example, you would have to take control over the bird and fly it around as it foraged. The ability to see, land on, or get nectar from the flower would be influenced by the genetic characteristics of the birds and flowers. My initial reaction was to dismiss the notion of mini-games as a distraction. Mini-games should not be the chocolate reward for consuming the nutritious broccoli making up the rest of the game. But as I think about it more in the context of virtual pets, as demonstrated by the other games in this genre, it may make a lot of sense. People bond strongly with their pets through play. Given the strong analogous patterns seen with the way people bond with virtual pets (Turkle 2005), some additional play would likely bring about additional bonding, and consequently increase motivation to improve strategies and keep the virtual pets alive.

After thinking this through and continuing the conversation on mini-games, the discussion took a turn for the worse. One student continued to pursue his ideas of in-game mini-games, stating "You know that game Oregon Trail and how you have to hunt the buffalo?" I thought it was ironic (and still sad) that we've come full circle back to the Oregon Trail. But maybe there is something positive to learn from the persistence of that game, and the fondness with which students remember it. Or maybe we just need better examples of educational games for them to recall.

# 13 Conclusion

Many critics of learning games oppose this medium for the ideas, concepts, and content that games can't teach well. Games may not be good at teaching you to factor polynomials or to memorize when "i" goes before "e." But whether they are "good" at teaching these ideas (absolutely or relative to other teaching modes) is irrelevant. There are likely quite a number of things that games are not good at teaching, and we should focus on the things that they can teach well. Few people actually argue that games should be the *only* teaching medium. Most learning games scholars and experts advocate for them being *some part* of an ecology of tools and methods that we use to teach and learn in schools, informal learning centers, and training situations. This use of games would be most effective if it were accompanied by changes in the pedagogies of learning institutions, to focus more on constructivist, learner-centered, motivating, and engaging tasks more generally.

## Mobile Game Affordances

Mobile games excel at connecting to existing classroom ecologies, and extending them in powerful new directions. These games fit naturally into the current landscape of teaching and learning in current schools, and also can connect school, game play, and the real world. This isn't to say that console- or PC-based video games can't or shouldn't be used for learning, but for many desired domains of learning, mobile games can fit more easily into the learning environment and connect more readily to learning outcomes. From the examples that we have implemented, and from the lessons of others, we have seen that mobile games have a number of affordances:

Game is in the eye of the beholder.   Mobile games don't have to look like traditional games. While effective mobile games must still draw upon good game design principles for both engaging and educating players, they don't need to advertise their "game" nature. If a teacher or administrator wants mobile game use not to *look* like a game, they can see it in this light and call it an "activity." But a student who wants to see it as a game (as most do) can easily do so.

Previous experience applies.   Directing activities based on mobile learning games feels much like running a multitude of other nontechnological activities such as role-play and simulation. Many teachers already feel comfortable facilitating activities in this way and, as we've seen, they can readily transfer these skills to facilitating classes based on mobile games.

Access.   While the student-to-computer ratio is rapidly decreasing we are still very far from one to one in most schools. Mobile devices can be easily and inexpensively deployed and maintained. As the capabilities of mobile devices increase (and reach some level of standardization), most people will have ready access to them, and these devices can be appropriated for learning.

Balance of onscreen and off-screen time.   It is relatively easy to move people involved in an activity employing mobile games on and off the screen rapidly. Games of the type I have described here don't demand 100 percent focus on the screen. In fact, in most cases players spend less than 20 percent of their time looking at the screen. So it is easy to create activities that balance that time with time spent in teacher- and student-led discussion as well as just about any other type of activity.

Temporal flexibility.   Mobile games can be started and stopped quickly, both from a technical perspective and from a design perspective. A class could play a mobile learning game for a few minutes a day over many classes or for hours at a time as the situation demands. This time need not be during actual class time, but could be throughout the day, week, or month. As will be discussed, this will likely lead to a situation in which mobile games can be integrated throughout the lives of learners in short bursts.

Natural communication.   Players using mobile learning games may have special communication capabilities enabled by the game, including instant messaging, bulletin boards, or even Voice over Internet Protocol (VoIP). But they can also use their face-to-face presence to communicate verbally, through gestures and facial expressions, and even through subtle body language. This freedom of expression enables effective communication in game and builds real communication skills out of game.

Learning is situated.   Mobile games can be set in social and physical contexts that readily become a part of the game. Even games that do not have technology-enabled multiplayer participation naturally incorporate the interactions of players with co-located players. Players can help each other, observe each other, and act together to create communities as they learn to solve problems. Through role play the communities in which they participate take on an even greater meaning.

Learning is embodied.   Players can not only communicate with their bodies but also think with their bodies using the physical context of games. It is easy to make physical connections with games set in the physical world.

Knowledge is constructed.   The role of the mobile device in these learning games is primarily to provide a "light" addition of virtual information into the interactions of the players. As such, there is little information fed to the players by their devices. Rather, they must create that knowledge and understanding through exploration.

Transfer is near.   Through real-world role play players experience a situation from perspectives that closely resembles real-life roles. We know that promoting transfer is difficult. Educational designers struggle with this challenge. Creating activities in which transfer doesn't have to go very far makes that task much easier.

## Tenets of Mobile Game Design

As mentioned earlier, not all learning is applicable to mobile games. Not all learning games are even applicable to mobile games. But there are many instances in which mobile learning games can be a tremendous asset and a good fit. As technologies create greater capacities and more ubiquitous use, while educators and developers push these games in new directions, those instances will grow exponentially. Through the work presented here we have developed a number of guiding tenets in choosing the uses of our mobile games. The development of these tenets is one of the reasons why game design is best seen as an iterative process aligning educational, technological, and interactive components. While these principles help us identify ripe opportunities for mobile learning games, we simultaneously consider what educational arenas are in need of technological support. Our guiding tenets:

Play close to reality.   Many of the games described in this book were "lightly" augmented reality games that were able to take advantage of many of the assets of the real world. Some explicitly incorporated the real world context in their design, and additional real world elements more subtly entered the game through the perception of players. Designing games that can take advantage of these real-world assets maximizes the potential of mobile games. This means imagining that the real world either represents itself exactly (e.g., the game is played at the nature center that the game is about) or something very similar (e.g., the game is played at the nature center that the game is about, but back in time fifty years, or it represents a similar nature center somewhere else).

Amplify authenticity.   In our move to make learning authentic we can do well to incorporate as much of authentic practice into games as possible. This means identifying the core practices associated with particular disciplines and making those an essential element of game play. Again, this can be done with light technological support, relying more heavily on human role play. The technology is best used to support the role-supplying tools, information, and perspectives associated with that role.

Create communities of practice.   The physical co-location of players enables easy and natural communication among the players in mobile games. Whether this co-location happens as a necessary structure over the course of an entire game, or serendipitously throughout a longer game, it allows players to come face to face to communicate and work together. The technology should foster the interactions among the players, facilitating their collaborative work and enabling them to learn from each other.

Incorporate location (or not).   Earlier I described two good uses of mobile technologies—those where location mattered and those where it did not. These are both applicable in design. Games that explicitly connect to and incorporate location take advantage of those unique assets. On the other hand, games that are designed to go anywhere can be readily incorporated in many different places from schoolyards to homes to workplaces to museums.

Bend time. Mobile games are ideal for playing in short spurts. This is supported by technology with instant or always-on capabilities, as well as devices that can be taken in and out of view quickly. This idea of playing in short bursts over long periods of time can enable games to be integrated with life, as shown in games such as Palmagotchi. Learners can play out of class and, through client-server technologies, can connect that experience back to the lessons in class.

Tie games to content, don't deliver it. Content is still key. It is hard to learn higher-order skills without content to ground them in, and that content can often be important to master in its own right. Mobile games can be tied explicitly to content, which can more readily be measured, while they are simultaneously building higher-order thinking skills.

Promote diverse forms of communication. Face-to-face communication is easy to incorporate into mobile learning games. It is important to build communication skills, and these games can foster expert communication in activities. More diverse forms of communication are becoming increasingly necessary, spanning a multitude of electronic media. These can be incorporated into mobile learning games.

Create deep casual experiences. Most of all, mobile learning games can be used to create learning experiences that are deep and powerful while staying casual. That is, these games do not need to be the kinds that only appeal to hard-core gamers. Instead they should be casual, flexible, and broadly appealing. But they should not be superficial. Flexibility in time, space, and context enables mobile learning games to fill both roles simultaneously.

**Twenty-first-Century Schools, Skills, and Tools**

There is undoubtedly a disconnect between students' experiences of media and technology in school and in their lives outside of school (Levin et al. 2002). Teens are avid consumers of Internet resources and media, and they know how to access, if not understand, vast amounts of information with ease. One could argue that given the rapidity with which students take up information with the aid of technology, we need not have academic focus on these tools. They will learn them outside of school, and in school we can focus on traditional learning, and get back to the basics that they are not building using these new media and technologies. But accessing information is not equivalent to understanding information. Instead of swimming in the deep invigorating waters of knowledge that have propelled us into the twenty-first century, today's students are drowning in a shallow sea of superficial understanding. We must provide better ways of helping them navigate this challenging intellectual landscape.

I am often asked how I can explain the facility with which students not only consume but also create content on the Internet, if they are truly struggling with these riches of information. A Pew study of teens as content producers and consumers (Lenhart and Madden 2005) showed that about half of American teens between ages twelve and seventeen are "Content Creators." The study defines Content Creators as reporting "having done one or more of the following activities: create a blog; create or work on a personal webpage; create or work on a webpage for school, a friend, or an organization; share original content such as artwork, photos, stories, or videos online; or remix content found online into a new creation."

If half of teens are Content Creators, then must we be doing something right? I argue that these statistics are not as promising as they first seem. First, the glass is half empty. Fully half of all teens have never created anything on the Internet. Not a Web page, story, or even a photo. Nearly half of the students in the survey didn't even know how to upload files to the Internet. While this is a simple learned skill, it is indicative of a culture of passive consumption. Clearly we are leaving quite a number of children behind. This number should be 100 percent at this point in time, as everyone should have been an Internet participant at some point in school. Second, the definition of "Content Creator" is fairly weak. This includes having ever done a single instance of one of these activities. A student who posts a picture as part of a school assignment is classified as a creator under this definition. Substantial active participation in the Internet is left undefined. True, nearly 20 percent of students reported having blogged, but these percentages get whittled down upon closer inspection.

Twenty-first-century students are certainly a part of Internet culture, and these levels of teen participation are increasing due to the prevalence of teen social networking sites. We should not take for granted, however, that nearly ubiquitous access is sufficient to prepare students for a future in which they must appropriate these technologies in diverse ways for life, learning, and work. Even as they become more immersed in this culture of technological participation, they still struggle with its application outside of simple uses in their personal lives. Schools need to play a significant role in better preparing students to apply these tools to understanding complex ideas. Transfer will not happen on its own, but these skills can grow and students can learn to apply them outside of their simple social context.

We cannot expect students on their own to jump the chasm from their current position of extensive casual use of information to the other side where participation is deep, meaningful, and built on understanding. We should not expect large jumps in understanding, even with expert guidance. Instead we must build bridges from where students are now to where we hope and expect this understanding to be. Handheld technologies can be the perfect guides to help them along those pathways. By creating deep, rich experiences that incorporate real-world challenges, social participation, and an array of media and information we can provide students with engaging activities that help them grow their own understanding. These lightweight activities situate students in learning environments that can truly grow these twenty-first-century skills.

We saw how activities such as the participatory simulations in chapter 6 *engaged students in sustained reasoning*. They *managed complexity* in the form of outbreaks of diseases in chapters 6 and 9 and the virtual ecologies of chapter 12. Students *tested solutions* in engineering in Environmental Detectives discussed in chapter 7 and solutions in epidemiology in Outbreak @ The Institute in chapter 10. Many times in those situations students were required to *manage problems in faulty solutions* as they worked with imperfect information and feedback in those same games shown in chapters 7, 8, and 10. They consistently had to *organize and navigate information structures and evaluate information*, though this was particularly prominent in POSIT, as shown in chapter 11. *Collaboration* was also ubiquitous and became more so over time as shown in the progression of chapter 10. *Communicating with other audiences* was also a focus of POSIT in chapter 11, but integrated into many other activities through presentations. Students learn to be prepared and to *expect the unexpected* as they managed complex systems such as those in Palmagotchi presented in chapter 12. Finally, the skills of *anticipating changing technologies* and *thinking about information technology abstractly* came from integrating not only handheld gaming technologies, but also audio, video, Wi-Fi, GPS, Bluetooth, Infrared, instant messaging, bulletin boards, and client-server technologies into many games. Students learned about the functionality of these technologies in fundamentally practical ways.

While education reform is pushing "back to the basics" in an attempt to provide students with a baseline of skills, it is missing the boat in delivering these critical twenty-first-century skills. Hopefully these mobile games can sneak these skills into the schools riding on the back of the content that the games also purport to teach.

### The Future of Mobile Learning Games

As mentioned earlier, mobile learning games are at their best when they are both deep and casual. This may seem like a paradox, as most casual games are fairly shallow, while most deep games are designed for hardcore gamers. But mobile technologies can change that through sustained experiences with only occasional participation. People can engage in mobile games for a few minutes (or even a few seconds) at a time, but distributed over hours, days, or even weeks. This provides for many opportunities to consider the game play thoughtfully, discuss it with others, and reflect on its significance, without requiring substantial investments in game-play time. This is the future of mobile learning games, games that take place any time and anywhere. As we saw in the instance of Palmagotchi, players were able to embed this game in the course of their daily routine and use class time to reflect on their learning. A similar experience was created around a multiday version of POSIT where players could incorporate interviews of virtual characters into their daily routine, going to see the NPC in the dining hall when they happened to be passing by. But being successful at these games means *thinking* about the game and corresponding strategies frequently, though *actual game play* may be infrequent. So where is the future of mobile learning games? Everywhere.

# References

Alters, B., and S. Alters. 2001. *Defending evolution: A guide to the creation/evolution controversy.* Boston: Jones & Bartlett.

Barab, S., S. Warren, and A. Ingram-Goble. 2006. Academic play spaces: Designing games for education. Paper presented at the Annual Conference of the American Educational Research Association, San Francisco, April.

Barab, S. A., K. Squire, and B. Dueber. 2000. Supporting authenticity through participatory learning. *Educational Technology Research and Development* 48, no. 2: 37–62.

Beech, H. 2000. Japan's weird science. *Time Magazine*, July 17. Available at ⟨http://www.time.com/time/magazine/article/0,9171,49435,00.html⟩ (accessed September 25, 2007).

Bishop, B., and C. Anderson. 1990. Student conceptions of natural selection and its role in evolution. *Journal of Research in Science Teaching* 27: 415–427.

Björk, S., J. Holopainen, P. Ljungstrand, and R. Mandryk, eds. 2002. Special issue on ubiquitous games of personal and ubiquitous computing. *Personal and Ubiquitous Computing* 6, no. 5–6: 358–361.

Bransford, J. D., A. Brown, and R. Cocking. 1999. *How people learn: Brain, mind, experience, and school.* Washington, DC: National Academies Press.

Bransford, J. D., and D. Schwartz. 1999. Rethinking transfer: A simple proposal with multiple implications. In *Review of Research in Education*, vol. 24, ed. A. Iran-Nejad and P. D. Pearson, 61–100. Washington, DC: American Educational Research Association.

Brown, A., and J. Campione. 1996. Psychological theory and the design of innovative learning environments: On procedures, principles, and systems. In *Innovations in learning: New environments for education*, ed. L. Schauble and R. Glaser, 289–325. Mahwah, NJ: Erlbaum.

Brown, J., A. Collins, and S. Duguid. 1989. Situated cognition and the culture of learning. *Educational Researcher* 18, no. 1: 32–42.

Bruner, Jerome. 1986. *Actual minds, possible worlds*. Cambridge, MA: Harvard University Press.

Cheok, A., S. Fong, K. Goh, X. Yang, W. Liu, and F. Farbiz. 2004. "Human Pacman: A mobile entertainment system with ubiquitous computing and tangible interaction over a wide outdoor area." *Personal and Ubiquitous Computing* 8, no. 2: 71–81.

Colella, V. 2000. Participatory simulations: Building collaborative understanding through immersive dynamic modeling. *Journal of the Learning Sciences* 9, no. 4: 471–500.

Csikszentmihalyi, M. 1990. *Flow: The psychology of optimal experience*. New York: Harper and Row.

Cuban, L. 2001. *Oversold and underused: Computers in the classroom*. Cambridge, MA: Harvard University Press.

Dede, C., B. Nelson, D. Ketelhut, J. Clarke, and C. Bowman. 2004. Design-based research strategies for studying situated learning in a multi-user virtual environment. Paper presented at the International Conference on the Learning Sciences, Los Angeles, CA, June.

de Souza e Silva, A., and G. Delacruz. 2006. Hybrid reality games reframed: Potential uses in educational contexts. *Games and Culture* 1: 231–251.

Dewey, J. 1938. *Experience and education*. New York: Simon & Schuster.

Ducheneaut, N., and R. Moore. 2006. More than just "XP": Learning social skills in multiplayer online games. *Interactive Technology and Smart Education* 2, no. 2: 89–100.

Engel Clough, E., and C. Wood-Robinson. 1985. How secondary students interpret instances of biological adaptation. *Journal of Biological Education* 19, no. 2: 125–130.

Epstein, J., and R. Axtell. 1996. *Growing artificial societies*. Cambridge, MA: MIT Press.

Eustace, K., G. Fellows, A. Bytheway, M. Lee, and L. Irving. 2004. The application of massively multiplayer online role playing games to collaborative learning and teaching practice in schools. *Proceedings of the 21st Australasian Society for Computers in Learning in Tertiary Education (ASCILITE) conference*, Perth, Western Australia, December, ed. R. Atkinson, C. McBeath, D. Jonas-Dwyer, and R. Phillips, 263. Nedlands, Western Australia: ASCILITE.

Facer, K., R. Joiner, D. Stanton, J. Reid, R. Hull, and D. Kirk. 2004. Savannah: Mobile gaming and learning? *Journal of Computer Assisted Learning* 20, no. 6: 399.

Federation of American Scientists. 2006. Report on the Summit on Educational Games. Washington, DC: Federation of American Scientists.

Friedman, T. L. 2005. *The world is flat: A brief history of the twenty-first century*. New York: Farrar, Straus, and Giroux.

Galarneau, L., and M. Zibit. 2006. Online games for 21st century skills. In *Games and simulations in online learning: Research and development frameworks*, ed. D. Gibson, C. Aldrich, and M. Prensky, 59–88. Hershey, PA: Information Science Publishing.

Gardner, D., Y. Larsen, W. Baker, A. Campbell, E. Crosby, E. Foster, Jr., and N. Francis. 1983. *A nation at risk: The imperative for educational reform*. (ED 226 006). Washington, DC: National Commission for Excellence in Education.

Gartner, Inc. 2005. As cited in Cell phones for the people. Business Week Online. November. Available at ⟨http://www.businessweek.com/magazine/content/05_45/b3958061.htm⟩ (accessed January 2006).

Gee, J. 2003. *What video games have to teach us about learning*. New York: Palgrave.

Gee, J. P. 2004. *Situated language and learning: A critique of traditional schooling*. New York: Routledge.

Gibson, B. 2006. *Mobile fun and games III* (3rd ed.). Whitepaper published by Juniper Research. Available at ⟨http://www.juniperresearch.com/shop/products/whitepaper/pdf/White_Paper_Games.pdf⟩ (accessed September 25, 2007).

Hirsch, E. 1996. *The schools we need and why we don't have them*. New York: Anchor Books.

Hofstadter, D. 1982. Nomic: A game that explores the reflexivity of law. *Scientific American* 246, no. 6: 16–28.

Hsi, S. 2003. A study of user experiences mediated by nomadic Web content in a museum. *Journal of Computer Assisted Learning* 19: 308–319.

Improbable Research. 2007. Winners of the Ig Nobel Prize. Available at ⟨http://www.improbable.com/ig-pastwinners.html#ig1997⟩ (accessed January 25, 2007).

Jakobsson, M., and T. Taylor. 2003. The Sopranos meet Everquest: Social networking in massively multiplayer online games. Available at ⟨http://hypertext.rmit.edu.au/dac/papers/Jakobsson.pdf⟩ (accessed on February 15, 2007).

Johnson, D., R. Johnson, and J. Holubec. 1994. *The new circles of learning: Cooperation in the classroom and school*. Alexandria, VA: Association for Supervision and Curriculum Development.

Johnson, S. 2005. *Everything bad is good for you: How today's popular culture is actually making us smarter*. New York: Riverhead Books.

Juul, J. 2003. The game, the player, the world: Looking for a heart of gameness. In *Level up: Digital games research conference proceedings*, ed. M. Copier and J. Raessens, 30–45. Utrecht: Utrecht University.

Kirriemuir, J., and A. McFarlane. 2004. *Report 8: Literature review in games and learning*. Bristol, UK: NESTA Futurelab.

Klemmer, S., B. Hartmann, and L. Takayama. 2006. How bodies matter: Five themes for interaction design. In *Proceedings of DIS 2006: ACM conference on the design of interactive systems*. State College, PA. New York: ACM Press.

Klopfer, E., and J. Groff. 2007. Accessing the dynamics of social learning using participatory simulations. Submitted to the conference on Computer Supported Collaborative Learning (CSCL 2007).

Klopfer, E., and K. Squire. 2007. Case study analysis of augmented reality simulations on handheld computers. *The Journal of the Learning Sciences* 16, no. 3: 371–413.

Klopfer, E., and K. Squire. 2003. Augmented reality on PDAs. Paper presented at the annual meeting of the American Educational Research Association, Chicago, IL, April.

Klopfer, E., K. Squire, and H. Jenkins. 2002. Environmental Detectives: PDAs as a window into a virtual simulated world. In *Proceedings of the international conference on wireless mobile technologies in education*. August, Vaxjö, Sweden. Los Alamitos, CA: IEEE.

Klopfer, E., and S. Yoon. 2005. Using palm technology in participatory simulations of complex systems: A new take on ubiquitous and accessible mobile computing. *Journal of Science Education and Technology* 14, no. 3: 287–295.

Klopfer, E., S. Yoon, and L. Rivas. 2004. Comparative analysis of palm and wearable computers for participatory simulations. *Journal of Computer Assisted Learning* 20: 347–359.

Kuhn, D. 1999. A developmental model of critical thinking. *Educational Researcher* 28, no. 2: 16–26, 46.

Laurel, B. 2001. *Utopian entrepreneur*. Cambridge, MA: MIT Press.

Lave, J., and S. Chaiklin, eds. 1993. *Understanding practice: Perspectives on activity and context*. Cambridge: Cambridge University Press.

Lave, J., and E. Wenger. 1991. *Situated learning: Legitimate peripheral participation*. Cambridge: Cambridge University Press.

Lee, K. 2004. Presence, explicated. *Communication Theory* 14: 27–50.

Lenhart, L., and M. Madden. 2005. Teen content creators and consumers: More than half of online teens have created content for the internet; and most teen downloaders think that getting free music files is easy to do. Pew Internet and American Life Project. Available at ⟨http://www.pewinternet.org/PPF/r/166/report_display.asp⟩ (accessed September 5, 2007).

Levin, D., S. Arafeh, A. Lenhart, and L. Rainie. 2002. The digital disconnect: The widening gap between Internet-savvy students and their schools. Pew Internet and American Life Project. Available at ⟨http://www.pewinternet.org/PPF/r/67/report_display.asp⟩ (accessed September 15, 2007).

Levy, F., R. Murnane. 2004. *The new division of labor: How computers are creating the next job market*. Princeton, NJ: Princeton University Press.

Lombard, M., and T. B. Ditton. 1997. At the heart of it all: The concept of presence. *Journal of Computer-Mediated Communication* 13, no. 3: n.p.

McGonigal, J. 2003. This is not a game: Immersive aesthetics and collective play. In *Proceedings of the Fifth International Digital Arts and Culture Conference*. RMIT, Melbourne, Australia. May 19–23, ed. A. Miles. Melbourne: Royal Melbourne Institute of Technology.

Magerkurth, C., A. Cheok, R. L. Mandryk, and T. Nilsen. 2005. Pervasive games: Bringing computer entertainment back to the real world. *ACM Computers in Entertainment* 3, no. 3: 4.

Mandryk, R., K. Inkpen, M. Bilezikjian, S. Klemmer, and J. Landay. 2001. Supporting children's collaboration across handheld computers. In extended abstracts of *CHI, Conference on Human Factors in Computing Systems*. Seattle, April. Available at ⟨http://dub.washington.edu/projects/geney/CHI-2001ShortSubmission.pdf⟩ (accessed September 25, 2007).

Massachusetts Department of Education. 2001. Massachusetts English language arts curriculum framework. Malden, MA: Massachusetts Department of Education.

McComas, W. 1996. Ten myths of science: Reexamining what we think we know about the nature of science. *School Science and Mathematics* 96: 10–16.

Means, B. 1994. Introduction: Using technology to advance educational goals. In *Technology and education reform: The reality behind the promise*, ed. B. Means, 12. San Francisco: Jossey-Bass.

Milgram, P., and F. Kishino. 1994. A taxonomy of mixed reality visual displays. *IEICE Transactions on Information Systems* E77-D, no. 12: 1321–1329.

Murnane, R., and F. Levy. 1996. *Teaching the new basic skills: Principles for educating children to thrive in a changing economy*. New York: Free Press.

National Center for Educational Statistics. 2006. Internet access in U.S. public schools and classrooms: 1994–2005. Available at ⟨http://nces.ed.gov/pubs2007/2007020.pdf⟩ (accessed September 15, 2007).

National Center on Education and the Economy. 2007. *Tough choices or tough times: The report of the new commission on the skills of the American workforce*. San Francisco: Jossey-Bass.

National Commission on Excellence in Education. 1983. *A nation at risk: The imperative for educational reform*. Washington, DC: U.S. Government Printing Office.

National Research Council. 1999. *Being fluent with information technology*. Washington, DC: National Research Council, National Academies Press.

National Research Council of the National Academies. 2006. *ICT fluency and high schools: A workshop summary.* Washington, DC: National Academies Press.

Nepf, H. 2002. Groundwater pollution curriculum package. Cambridge, MA: Center for Environmental Health Sciences. Available at ⟨http://web.mit.edu/nepf/www/guide.html⟩ (accessed on January 15, 2007).

Piaget, J. 1977. *The development of thought: Equilibration of cognitive structures,* trans. A. Rosin. New York: Viking Press.

Piekarski, W., and B. Thomas. 2002. ARQuake: The outdoor augmented reality gaming system, *Communications of the ACM* 45, no. 1: 36–38.

Prensky, M. 2001. *Digital game-based learning.* New York: McGraw Hill.

Rambusch, J., and T. Ziemke. 2005. The role of embodiment in situated learning. In *Proceedings of the 27th annual conference of the Cognitive Science Society,* ed. B. G. Bara, L. Barsalou and M. Bucciarelli, 1803–1808. Mahwah, NJ: Erlbaum.

Ravitch, D. 2000. *Left back.* New York: Touchstone.

Reuters. 2005. Handhelds fuel 21 pct rise in U.S. video game sales. Reuters, July 28, 2005. Available at ⟨http://gameinfowire.com/news.asp?nid=6752⟩ (accessed December 19, 2007).

Reuters. 2006. Tamagotchis seek second wave of virtual pet owners. Reuters, April 20, 2006. Available at ⟨http://www.boston.com/business/technology/articles/2006/04/20/tamagotchis_seek_second_wave_of_virtual_pet_owners/⟩ (accessed September 25, 2007).

Salen, K., and E. Zimmerman. 2003. *Rules of play: Game design fundamentals.* Cambridge, MA: MIT Press.

Salzman, M., C. Dede, R. Loftin, and J. Chen. 1999. A model for understanding how virtual reality aids complex conceptual learning. *Presence: Teleoperators and Virtual Environments* 8, no. 3: 293–316.

Sandford, E., M. Ulicsak, K. Facer, and T. Rudd. 2006. Teaching with games: A one-year project supported by Electronic Arts, Microsoft, Take-Two and ISFE. In *Final report: Using commercial off-the-shelf computer games in formal education.* Bristol, UK: NESTA Futurelab.

Sandoval, W. 2005. Understanding students' practical epistemologies and their influence on learning through inquiry. *Science Education* 89: 634–656.

Scardamalia, M., and C. Bereiter. 1991. Higher levels of agency for children in knowledge building: A challenge for the design of new knowledge media. *The Journal of the Learning Sciences* 1, no. 1: 37–68.

Schwabe, G., and C. Göth. 2005. Mobile learning with a mobile game: Design and motivational effects. *Journal of Computer Assisted Learning* 21: 204–216.

Schön, D. 1983. *The reflective practitioner.* New York: Basic Books.

Schön, D. 1987. *Educating the reflective practitioner.* San Francisco: Jossey-Bass.

Shaffer, D. 2006. How computer games help children learn. New York: Palgrave Macmillan.

Shaffer, D., and M. Resnick. 1999. Thick authenticity: New media and authentic learning. *Journal of Interactive Learning Research* 10, no. 2: 195–215.

Shin, N., C. Norris, and E. Soloway. 2006. Effects of handheld games on students learning in mathematics. In *Proceedings of the 7th International Conference on Learning Sciences,* Bloomington, IN, June 27–July 1, ed. By S. A. Barab, K. E. Hay, and D. T. Hickey, 702–708. New York: Routledge.

Soloway, E., C. Norris, P. Blumenfeld, B. Fishman, J. Krajcik, and R. Marx. 2001. Devices are ready at hand. *Communications of the Association for Computing Machinery* 44, no. 6: 15–20.

Squire, K., and M. F. Jan. 2007. Mad city mystery: Developing scientific argumentation skills with a place-based augmented reality game on handheld computers. *Journal of Science Education and Technology* 16, no. 1: 5–29.

Squire, K., and H. Jenkins. 2003. Harnessing the power of games in education. *InSight* 3, no. 1: 7–33.

Steinkuehler, C. 2004. Learning in massively multiplayer online games. In *Proceedings of the Sixth International Conference of the Learning Sciences,* ed. Y. B. Kafai, W. A. Sandoval, N. Enyedy, S. A. Nixon, and F. Herrera, 521–528. Mahwah, NJ: Erlbaum.

Steinkuehler, C. A. 2006. Games as a highly visible medium for the study of distributed, situated cognition. In *Proceedings of the 7th International Conference on Learning Sciences,* Bloomington, IN, June 27–July 1, ed. By S. A. Barab, K. E. Hay, and D. T. Hickey, 1048–1049. New York: Routledge.

Strijbos, J.-W., P. A. Kirschner, and R. L. Martens, eds. 2004. *What we know about CSCL: And implementing in higher education.* Norwell, MA: Kluwer.

Time Asia. 1998. Time Cyberelite: 41—Aki Maita. October 12. Available at ⟨http://www.time.com/time/asia/asia/magazine/1998/981012/toys_r_her.html⟩ (accessed January 25, 2007).

Tinker, R., and J. Krajcik, eds. 2001. *Portable technologies: Science learning in context.* New York: Kluwer Academic/Plenum.

Turkle, S. 1995. *Life on the screen: Identity in the age of the internet.* New York: Simon & Schuster.

Turkle, S. 2005. Computer games as evocative objects: From projective screens to relational artifacts. In *Handbook of computer game studies*, ed. J. Raessens and J. Goldstein, 267–282. Cambridge, MA: MIT Press.

Tyack, D., and L. Cuban. 1997. *Tinkering towards utopia: A century of public school reform*. Cambridge, MA: Harvard University Press.

Tyack, D., and W. Tobin. 1994. The "grammar" of schooling: Why has it been so hard to change? *American Educational Research Journal* 31, no. 3: 453–479.

von Glasersfeld, E. 1995. A constructivist approach to teaching. In *Constructivism in education*, ed. L. P. Steffe and J. Gale, 3–15. Hillsdale, NJ: Erlbaum.

Vygotsky, L. 1978. *Mind in society: The development of higher psychological processes*. Cambridge, MA: Harvard University Press.

Walz, S. 2005. Constituents of hybrid reality: Cultural anthropological elaborations and a serious game design experiment merging mobility, media, and computing. In *Total interaction. Theory and practice of a new paradigm for the design disciplines*, ed. G. M. Buurman, 122–141. Basel: Birkhäuser.

Wenger, E. 1998. *Communities of practice. learning, meaning, and identity*. Cambridge: Cambridge University Press.

Wenglinsky, H. 1998. *Does it compute? The relationship between educational technology and student achievement in mathematics*. Princeton, NJ: Educational Testing Service, Policy Information Center, Research Division.

Wilensky, U., and W. Stroup. 1999. Learning through participatory simulations: Network-based design for systems learning in classrooms. In *Proceedings of the Computer Support for Collaborative Learning*, Stanford University, December 12–15, ed. C. Hoadley and J. Roschelle, 80. Mahwah, NJ: Erlbaum.

Winn, W. D. 2003. Learning in artificial environments: Embodiment, embeddedness, and dynamic adaptation. *Technology, Instruction, Cognition, and Learning* 1, no. 1: 87–114.

Winn, W. D., and M. Windschitl. 2001. Learning in artificial environments. *Cybernetics and Human Knowing* 8, no. 4: 5–23.

Winn, W., M. Windschitl, R. Fruland, N. Hedley, and L. Postner. 2001. *Learning science in immersive virtual environment*. Paper presented at the annual meeting of the American Educational Research Association, Seattle, WA, April.

Winn, W., M. Windschitl, R. Fruland, and Y. Lee. 2002. When does immersion in a virtual environment help students construct understanding? In *Proceedings of the International Conference on the Learning Sciences*, ed. P. Bell and R. Stevens. Mawah, NJ: Erlbaum.

Wudunn, S. 1997. Hatchling of pet lover is the rage of toylands. *New York Times*, September 7.

Yee, N. 2006. The demographics, motivations and derived experiences of users of massively multiuser online graphical environments. *PRESENCE: Teleoperators and Virtual Environments* 15: 309–329.

Yoon, S. (forthcoming). An evolutionary approach to harnessing complex systems thinking in the science and technology classroom. To appear in the *International Journal of Science Education*.

# Index